SpringerBriefs in Ecology

For further volumes:
http://www.springer.com/series/10157

Robert E. Loeb

Old Growth Urban Forests

Springer

Robert E. Loeb
The Pennsylvania State University
College Place
DuBois, PA 15801, USA
rxl5@psu.edu

Additional material to this book can be downloaded from http://extras.springer.com

ISSN 2192-4759   e-ISSN 2192-4767
ISBN 978-1-4614-0582-5   e-ISBN 978-1-4614-0583-2
DOI 10.1007/978-1-4614-0583-2
Springer New York Dordrecht Heidelberg London

Library of Congress Control Number: 2011932490

Printed on acid-free paper

Springer is part of Springer Science+Business Media (www.springer.com)

Dedicated to my dear wife Susan, son Steven, and granddaughter Morgan for their inspiration of my research on old growth urban forests.

Preface

My purpose in writing this book is to motivate urban foresters and ecologists to break through a barrier created by the hallowed concept of old growth forest as "undisturbed by humans." Many urban forests are composed of old trees and admired by visitors as being old growth forests. However, the traditional scientific education afforded foresters and ecologists rejects classification of any urban forest as old growth because the forest did not achieve the successional climax state as a result of human disturbances. Research on urban forests is not hindered by an artificial conceptual framework of age in rural forests. Recognizing old growth forests in urban settings expands the spectrum of urban forest research to include a focus on long-term changes in the relationships between human-caused changes and urban forest dynamics. Even foresters and ecologists conducting research in rural old growth forests will reap the benefits of understanding critical issues such as species introductions and climate change that have affected old growth urban forests across our planet for centuries.

Because an unfamiliar genre is best exemplified by specific examples from a real-world context, I provide old growth original research from the forest region and metropolitan areas I have studied for decades. A critical analysis of the research is provided to orient current and future urban foresters and ecologists as well as advocates for the forests to the historical ecology methods employed to conduct research in old growth urban forest dynamics. This research is both theoretical and applied because the forests are typically found on public lands and have already gained public recognition. Consequently, this book also provides a science-based framework for the management and restoration of old growth urban forests through partnerships of foresters, scientists, educators, students, government officials, and community neighbors. In the final view, only joint efforts will move the science forward and permit old growth urban forests to remain a part of human communities for centuries into the future.

DuBois, PA, USA   Robert E. Loeb

Acknowledgments

During my teen years growing up in The Bronx, John Reed of the New York Botanical Gardens employed me as a research assistant in several scientific projects for which I am very grateful, but most important to my research on old growth urban forests were the many valuable lessons I learned while remapping all of the trees and interviewing visitors to the 12 ha old growth Hemlock Forest. My thanks are extended to my undergraduate research mentor Jon Greenlaw for introducing me to the relationships between animals and forests, to my graduate research mentor Calvin Heusser for expanding my research methods to include Quaternary palynology, and to my research collaborators George Profous and Rowan Rowntree for enlarging my perspective on urban forests to encompass the whole world. My brother Francis and sister-in-law Elizabeth deserve special thanks for spending many long hours on frozen lakes developing an effective freezing corer to obtain unconsolidated recent sediments in stratigraphic order for paleopalynological research.

During the years spanning my undergraduate and graduate studies I was an advocate for community involvement with old growth urban forests which soon led to my public service as the Parks and Recreation Committee Chair of Community Planning Board 12, The Bronx. I am grateful to many elected and appointed officials, public servants, and citizens for helping me to understand the spectrum of views on urban forests and recreation as well as a variety of ways to bring people together to move forward old growth urban forest management and restoration in a time of municipal financial crisis.

The support of the Fairmount Park Commission, especially Thomas Witmer to conduct research in Fairmount Park and Robert Armstrong to access archival documents, is deeply appreciated. Adam Levine provided an invaluable service in sharing the digitized maps of Fairmount Park. My thanks are extended to Peter Kroll of the Haddon Township Environmental Commission as well as Janet Goehner-Jacobs and Robyn Jeney of the Saddler's Woods Conservation Association for giving their approval to conduct the research in Saddler's Woods, New Jersey and sharing their experiences with invasive species management. Frank Clements, Park Manager, and Marc Muller, Resident Engineer of the National Zoological Park, Washington DC, provided valuable information and support for the research in Walbridge Woods. I appreciate the funds made available by The Pennsylvania State University for a sabbatical to conduct the field work in Fairmount Park and research development grants for the sampling in Saddler's Woods and Walbridge Forest. My thanks are extended to John Gibson, Robert Eck, Tami Jones, Robin Sterling, Jay Hartman, Rosemary Nichols, and William Thomas of the Tree Planting and Preservation Committee, Moorestown Township, New Jersey, for sharing their tree inventory information. Thanks are extended for the efforts of Bryant Cowart and A. Knadya O'Kelly of the Clean and Green Environmental Education Center, City of East Orange, New Jersey, to furnish tree inventory data. The tree survey report given to me by the members of the Shade Tree Commission of Haddonfield, New Jersey, William Polise, Robin Potter, Jeff Hammon, Marjorie Coar, Harriet Monshaw, and Anne Koelling, is appreciated. Alexander McCartney of the New Jersey Community Forestry Program imparted valuable guidance concerning street forests in New Jersey. Information provided by Kenneth Ferebee, Diane Pavek, and John Schmit of the National Park Service is appreciated. I appreciate the figures showing the need for forest restoration provided by Mark Mead of the Green Seattle Partnership – City of Seattle. The photograph of the new South Cove Trail, Radnor Lake State Natural Area, Nashville, Tennessee, furnished by Park Ranger Joshua Walsh is valued. Many thanks to Steve Ward, Park Manager, Radnor Lake State Natural Area for long, thoughtful conversations that have recalibrated my perspective on old growth urban forest management.

Contents

Chapter 1
What and Where Are Old Growth Urban Forests?

Abstract Old growth urban forests have developed in cities around the earth with human actions that have changed arboreal composition and forest dynamics. For the term old growth urban forest, the concept "urban forest" refers to forests in a metropolitan area and "old growth" indicates forest development after a regional forest resetting event such as a war. Urban forestry and ecology literature in English revealed studies of street, landscaped, and remnant old growth urban forests in 28 countries and 62 metropolitan areas. Long-term changes in old growth urban forest structure are primarily determined by human activities causing species losses and arboreal population decimation. Most changes were intended, especially in regard to shifting fashions in species selection for arboricultural plantings; however, new species introductions are frequently not successful over the long term. The unintended or at least unforeseen effects of diseases and invasive species introductions often have caused the most devastating transmutations of old growth urban forests. Reversing impending losses in old growth urban forests that are unable to reproduce because of human modifications of the forest environment requires historical ecology research to determine the species composition and environmental conditions for the historical continuity of the forests.

Keywords Old growth urban forest • Historical continuity • Metropolitan areas • Forest resetting event • Arboricultural plantings • Street forest • Landscaped forest • Remnant forest

The common perception of old growth forest is a canopy composed of trees that have attained the maximum age of the various species. Arboreal species diversity in the subcanopy, sapling, and seedling layers is limited by the competition for light from canopy gaps and losses to browsing animals. Without human disturbance,

R.E. Loeb, *Old Growth Urban Forests*, SpringerBriefs in Ecology,
DOI 10.1007/978-1-4614-0583-2_1, © Springer Science+Business Media, LLC 2011

species composition can be related to the prevailing climate, soils, and topography. As well, forest dynamics can be associated with the effects of browsing animals and disturbances from storms including fires from lightning (Leverett 1996). Of course, an old growth forest researcher cannot be certain if people have affected species composition or forest dynamics (Cho and Boerner 1991; Ward et al. 1996).

In contrast, an old growth urban forest researcher can be certain that humans have affected species composition and forest dynamics. To understand the effect of this critical difference, there is a need to define the term old growth urban forest and to develop a forest typology. The first step is to consider a brief overview of urban forest history. Lawrence (2006) described the long history of urban forest types influenced by European cultures as a progression of forms. Trees planted in private gardens after the collapse of the Roman Empire is the first form of the urban forest. Plantations in streets began in the sixteenth century and followed the arrangement of two parallel lines of trees used in gardens. Planting trees in squares, plazas, and small parks, again mimicking private gardens' patterns, was introduced in the seventeenth century. During the eighteenth century, estates on the borders of the city, which encompassed much larger tracts than available in the cities, were landscaped with trees. The designs for estates influenced the landscaping of the large public parks that were created in response to public health concerns arising from the shift of the industrial revolution to cities in the nineteenth century. Urban expansion caused remnant forests to become enveloped within the ever-enlarging boundaries of cities since medieval times (Forrest and Konijnendijk 2005). Typically, remnant forests sustained tree harvests as well as deer and livestock browsing of the saplings and pollard trees (Rackham 2003). As European colonies were established in other continents, new arrangements for the distribution of urban forests arose with the opportunities to create cities. For example, the national government plan for frontier towns in Australia and New Zealand specifically included a belt of parkland separating the urban core from the surrounding suburban area which was the precursor of the greenbelt concept (Williams 1966). When travel in China and Japan was permitted, Europeans discovered unfamiliar styles of planting trees to form urban forests (Profous 1992; Cheng et al. 2000; Chen and Nakama 2010).

Based on history, the old growth urban forest typology contains three types: street, landscaped, and remnant (Fig. 1.1). The semi-linear layout of street forests is easily associated with their namesake location of streets but also fits planting patterns in gardens, cemeteries, memorial groves, and promenades. Landscaped forests originated in former estates and royal or government lands and the trees were planted following a scheme, which may have been recorded. Remnant forests can be divided into two primary origins: forest with limited tree harvesting and forest reestablished on abandoned agricultural fields or timber lands. The classification of a particular old growth urban forest into one of the three types is not always a clear decision because the site history may reveal that sections of the forest fit all three types. For example, cemeteries may contain tree-lined roadways, memorial groves, areas designed by many famous landscape architects, remnants of agricultural origin, and

Fig. 1.1 Old growth
urban forest types and forms

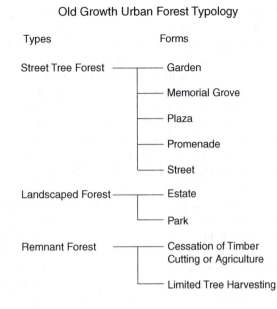

Fig. 1.1 Old growth urban forest types and forms

a valley forest with minor tree felling (Leland and Smith 1941; Denne 1987; Leighton 1987; Worpole 2003).

The brief overview of urban forest history may have caused some uncertainty as to the meaning of the term "urban forest" because forests outside of the city border are considered urban forests. Urban refers to a metropolis instead of just one city alone because a metropolis develops over time by one city dominating not only its surrounding countryside but also other cities and their countrysides (Careless 1954). Urban forests in a metropolis can be the street forest located in the central core of one city, the landscaped forest in the periphery of another city, and the remnant forest in a suburb between cities. Metropolitan areas have a long history of human influences on the constituent locales which enable comparisons among urban forests. The common history of human-caused changes creates equality among the three old growth urban forest types which must not be lost to the wide variety of terminology used in urban forestry (Konijnendijk et al. 2006). Therefore, the word "forest" is used whether reference is made to a single old growth urban site or every location in a metropolis.

In the term "old growth urban forest," the concept of "old growth" does not refer to the origin of the forest prior to historical records or a date greater than the maximum age of the tree species composing the forest. Instead, the determination of whether an urban forest can be classified as "old growth" is based on how the development of the particular forest is related to the most recent forest resetting event during the history of the metropolis. If the growth, reestablishment, or planting of a forest started before or soon after the forest resetting event, then the forest is considered old growth. The following are examples of forest resetting events: forest clearance for city formation (Popoola and Ajewole 2001), transformation from agricultural to urban land use (Loeb 1992a), war (Cheng and McBride 2006; Lacan and McBride 2009), forest leveling storms (Hytteborn and Packham 1985), and the transition from wood to fossil fuels as the primary source of energy (Tillman 1978; Antrop 2005). The key point is a researcher must decide which historical change resets the forests of the entire modern-day metropolis. The subsequent forest transitions (diseases – Solotaroff 1912; Hepting 1974, 1977; Tello et al. 1985; Rodoni et al. 1999; insect infestations – Solotaroff 1912; Dunlap 1980; Loeb 1989; fire – Loeb 2001; harvesting wood for income – Hoglund 1962; Bischoff 1994; and storms – Natti 1979) are incorporated into the explanation of changes in forest dynamics over time.

The historical ecology approach to old growth urban forest research provides a more thorough understanding of forest dynamics. The importance of forest dynamics to urban forestry in the United States was described by Rowntree (1998) as "Everyone wants to understand change – the way things were and the way they will be tomorrow." In a Council of Europe publication on urban forest management, Sukopp and Werner (1982) put forth "the principle of historical continuity" and wrote that "Primary habitats or those with a long history of the same kind of management are particularly valuable for nature conservation and should be identified and protected at all costs." The application of historical ecology research methods to determine long-term forest dynamics can be described as examining long-term changes in forest structure. However, this description is problematic because urban forest research measures forest structure in two ways: (1) sampling of tree population, location, size, and condition; and (2) ascertaining the spatial arrangement of vegetation in regard to buildings and roads (Nowak 1994). Although knowing where old growth urban forests are located within a metropolis is valuable information, historical urban forest ecology research is focused on changes through time in the canopy, subcanopy, sapling, and seedling layers in order to determine forest dynamics.

The primary goal of research, restoration, and management for old growth urban forests is the same regardless of the forest type – historical continuity. Comparative research on old growth urban forests in a metropolis is similar to applying the concept of an urban-rural gradient (Zipperer and Guntenspergen 2009) but the metropolis concept adds the complexity of historical forest development and change. A point of concern arises when a metropolitan area spans more than one forest region because differences in climate, soils, and topography become significant factors when comparisons are made among old growth urban forests. Considering the importance of human activities to old growth urban forest dynamics and the potential for differing environmental conditions among metropolitan areas, Gleason's (1939) Individualistic Concept is preferable to Clements' (1916) Theory of Succession as the theoretical basis for old growth urban forest comparisons (Rackham 2003).

   The historical ecology of a particular old growth urban forest is not totally independent of other old growth urban forests because common patterns of forest transformations can be found within and among metropolitan areas of a nation. For example, Rackham (1980) provided a comprehensive review of the historical ecology of ancient woodlands in England which has informed research in particular old growth urban forests (Cole and Mullard 1982; Layton 1985; Peglar et al. 1989; Skeggs 1999). A key common factor affecting old growth urban forests is historical changes in tree species plantings. The native species of a region were primarily planted until species introductions from the European colonies spread across the globe (Lawrence 2006). As with many other human decisions, fashion preferences and commercial availability controlled species selection in several countries (Australia – Spencer 1986; Banks and Brack 2003; Bennett 2010; Barbados – Watts 1963; China – Jim 2002; England – Jarvis 1973; Soutar and Peterken 1989; Dehnen-Schmutz et al. 2007; Mayhew 2009; Hong Kong – Jim 1992; Italy – Attorre et al. 2000; New Zealand – Shepherd and Cook 1988; Portugal – Bunting and Rego 1988; South Africa – Shaughnessy 1987; Spain – Austrich 1987; United States – Leighton 1970, 1986, 1987; Stearn 1972; Stein and Moxley 1992; Fry 1996; McPherson and Luttinger 1998; Adams 2004; Loeb 2010). Although an understanding of climate similarities among regions of the earth guided decisions as to species suitability for plantings, determining the fitness of a species was often a matter of testing in each locale. Typically, arboricultural introductions increased urban forest species diversity at first but species diversity dropped as specimens died. The problematic legacy of nonnative species introductions is the spread of invasive species, which has had the unintended consequence of displacing native species in old growth urban forests (Loeb 2010; Loeb et al. 2010).

   Urban forest research in metropolitan areas of many nations indicates there is great potential for comparative research on old growth urban forests. The listing of publications in Table 1.1 should be considered the minimum breadth of possibilities to begin comparative research because inclusion was based on a review of the scientific literature written in English. Comparative studies of old growth urban forests in the metropolitan areas noted in Table 1.1 will require in-depth examination of more locales including representatives of all three forest types – street, landscaped, and remnant. When the historical ecology of several old growth urban forests is known for a metropolis, then long-term trends in forest dynamics can be related to regional biotic factors including diseases and browsing animals and abiotic factors such as air pollution and the urban heat island (McCarroll and Loader 2004). Combining analysis results across metropolitan areas can provide answers for planetary climate change questions because old growth urban forests developed with local climate changes for more than a century (Carreiro and Tripler 2005).

Table 1.1 Global list of old growth urban forest research in English

Country	Region	Metropolis	Urban forest type	Reference
Australia	Cumberland	Sydney	Street, landscaped, remnant	Benson and Jowell (1990); Frawley (2010)
	New South Wales	Canberra	Street, landscaped	Banks and Brack (2003)
	South Australia	Adelaide	Landscaped	Peter (2008)
	Victoria	Melbourne	Street, landscaped	Prescott (1982); Yau (1982); Aldous (1990)
Brazil	São Paulo	São Paulo	Remnant	Pinheiro et al. (2006)
Canada	Ontario	Ottawa	Street	Dean (2005)
		Toronto	Street, landscaped, remnant	Andresen and Granger (1986)
China		Beijing	Street, landscaped, remnant	Profous (1992); Li et al. (2006)
		Guangzhou City	Street, landscaped	Jim and Liu (2001)
		Hong Kong	Street	Jim (1986)
Costa Rica		San Jose	Street	Monge-Nájera and Pérez-Gómez (2010)
Czech Republic		Prague	Street, landscaped	Profous and Rowntree (1993)
England	Berkshire	Reading	Remnant	Skeggs (1999)
	East Anglia	Diss	Street, landscaped	Peglar et al. (1989)
		London	Street, landscaped, remnant	Saunders (1969); Baker et al. (1978); Rackham (1978); Cole and Mullard (1982); Hanson (1983); Layton (1985); Jones (1993)
	Oxfordshire	Oxford	Remnant	Savill et al. (2010)
Finland		Helsinki	Street, landscaped, remnant	Seppä (1997)
Germany		Berlin	Street, landscaped, remnant	Brande et al. (1990)
		Several	Street, landscaped	Kunick (1987)
Greece		Athens	Street, landscaped	Profous et al. (1988)
India		Bangalore	Street, landscaped	Sudha and Ravindranath (2000); Nagendra and Gopala (2010)

Country	Region	City	Type	Reference
Ireland	Kerry	Killarney	Landscaped, remnant	Kelly (1981); Whaley (1985)
	Leinster	Dublin	Street, landscaped	Boylan (1992)
Iran		Tehran	Street, landscaped, remnant	McBride and Mossadegh (2000)
Italy		Florence	Street	Grossoni et al. (1999)
Japan		Okinawa	Street	Chen and Nakama (2010)
		Tokyo	Street, landscaped, remnant	Numata (1977); Cheng et al. (2000)
Kyrgyzstan	Chuy	Bishkek	Street	Whitehead (1982)
Malaysia		Kuala Lumpur	Street, landscaped, remnant	Webb (1998b)
New Zealand		Auckland	Remnant	Esler (1991)
		Christchurch	Street, landscaped	Stewart et al. (2009)
		Palmerston	Landscaped	Aldous (1981)
		Wellington	Landscaped, remnant	Bagnall (1979); Shepherd and Cook (1988); Poole (1992)
Nigeria	Oyo	Ibadan	Remnant	Popoola and Ajewole (2001)
Poland		Warsaw	Remnant	Solińska-Górnicka and Symonides (1990)
Scotland		Glasglow	Remnant	Dougall and Dickson (1997)
Singapore		Singapore	Street, landscaped, remnant	Chin and Corlett (1986); Webb (1998a)
South Africa	KwaZulu-Natal	Gauteng	Remnant	Grobler et al. (2002)
		Pietermaritzburg	Street, landscaped	Haigh (1984); Showers (2010)
Sweden		Stockholm	Remnant	Florgård (2000)
		Uppsala	Remnant	Hytteborn and Packham (1985)
Switzerland		Several	Landscaped	Lawrenz (1989)
		Geneva	Street, landscaped	Beer (1996)
Thailand		Bangkok	Street, landscaped	Thaiutsa et al. (2008)
Turkey		Ankara	Landscaped	Oguz (2004)

(continued)

Table 1.1 (continued)

Country	Region	Metropolis	Urban forest type	Reference
United States	California	San Francisco	Landscaped, remnant	McBride and Froehlich (1984); McBride and Jacobs (1986), Young (2004); McCubbin (2009)
		Lake Tahoe	Remnant	McBride and Jacobs (1986)
	Florida	Miami	Remnant	Ball (1980)
	Illinois	Chicago	Street, landscaped, remnant	Schmid (1975); Bowles et al. (2005)
		Urbana	Street	Dawson and Khawaja (1985)
	Indiana	Lafayette	Remnant	Bramble (1975)
	Massachusetts	Boston	Street, remnant	Welch (1994),
	New York	New York	Street, landscaped, remnant	Loeb (1982, 1992a, b, 2010); Stalter and Kincaid (2008)
		Syracuse	Street, landscaped, remnant	Sanders (1981); Nowak and O'Connor (2001); Zipperer (2002); Zipperer and Zipperer (2002)
	Ohio	Cleveland	Remnant	Loeb (2001)
		Columbus	Landscaped	Quigley (2002)
		Dayton	Street	Canfield and Runkle (1999)
	Oregon	Portland	Remnant	Poracsky and Houck (1994)
	Pennsylvania	Philadelphia	Landscaped	Loeb (2010)
	Tennessee	Memphis	Landscaped, remnant	Guldin et al. (1990); Heineke (2009)
		Nashville	Landscaped, remnant	Loeb et al. (2010)
	Washington	Seattle	Street, landscaped, remnant	Hanson et al. (2005)
	Wisconsin	Milwaukee	Street	Dorney et al. (1984)

References

Adams DW (2004) Restoring American gardens. Timber Press, Portland, Or

Aldous DE (1981) Native and exotic trees of the Esplanade, Palmerston North, New Zealand. Arboric J 5:291–300

Aldous DE (1990) Trees of the historic Burnley Gardens, Melbourne, Australia. Arboric J 14:61–73

Andresen JW, Granger WB (1986) Metropolitan Toronto's urban forest: history and future. Arboric J 10:309–318

Antrop M (2005) Why landscapes of the past are important for the future. Landsc Urb Plan doi:10.1016/j.landurbplan.2003.10.002

Attorre F, Bruno M, Francesconi F, Valenti R, Bruno F (2000) Landscape changes of Rome through tree-lined roads. Landsc Urb Plan doi:10.1016/S0169-2046(00)00069-4

Austrich RR (1987) El Real Jardin Botanico de Madrid and the glorious history of botany in Spain. Arnoldia 47:3–20

Bagnall RG (1979) A study of human impact on an urban forest remnant: Redwood Bush, Tawa, near Wellington, New Zealand. New Zealand J Bot 17:117–126

Baker CA, Moxey PA, Oxford PM (1978) Woodland continuity and change in Epping Forest. Field Stud 4:646–669

Ball MC (1980) Patterns of secondary succession in a mangrove forest of southern Florida. Oecologia doi:10.1007/BF00572684

Banks JCG, Brack CL (2003) Canberra's urban forest: evolution and planning for future landscapes. Urb For Urb Green doi:10.1078/1618-8667-00015

Beer R (1996) The role of trees in the urban environment: the example of Geneva. Arboric J 20:437–444

Bennett BM (2010) A global history of Australian trees. J His Biol doi:10.1007/s10739-010-9243-7

Benson DH, Jowell J (1990) Sydney's vegetation 1788–1988: utilization, degradation and rehabilitation. Proc Ecol Soc Aust 16:115–127

Bischoff A (1994) Rethinking the urban park: rediscovering urban solutions. In: Platt RH, Rowntree RA, Muick PC (eds) The ecological city: preserving and restoring urban biodiversity. Univ Mass Press, Amherst, Ma

Bowles M, Jones M, McBride J, Bell T, Dunn C (2005) Temporal instability in Chicago's upland old growth forests. Chicago Wilderness J 3(2):5–16

Boylan C (1992) Trees of Dublin. Arboric J 16:327–341

Bramble WC (1975) The story of McCormick Woods. Am For 81:16–21

Brande A, Böcker R, Graf A (1990) Changes of flora, vegetation and urban biotopes in Berlin (west). In: Sukopp H, Hejný S (eds) Urban ecology. Plants and plant communities in urban environments. SPB Academic Publishing, The Hague, Neth

Bunting SC, Rego FC (1988) Human impact on Portugal's vegetation. Rangel 10:251–255

Canfield B, Runkle JR (1999) Size structure and composition of trees in Oakwood, Ohio: historical and environmental determinants. Ohio J Sci 99(5) 102–110

Careless JMS (1954) Frontierism, metropolitanism and Canadian history. Can Hist Rev 35:1–21

Carreiro MM, Tripler CE (2005) Forest remnants along urban-rural gradients: examining their potential for global change research. Ecosyst doi:10.1007/s10021-003-0172-6

Chen B, Nakama Y (2010) A study on village forest landscape in small island topography in Okinawa, Japan. Urb For Urb Green doi:10.1016/j.ufug.2009.12.004

Cheng S, McBride JR (2006) Restoration of the urban forests of Tokyo and Hiroshima following World War II. Urb For Urb Green doi:10.1016/j.ufug.2006.07.003

Cheng S, McBride JR, Fukunari K (2000) The urban forest of Tokyo. Arboric J 23:379–392

Chin WY, Corlett R (1986) The city and the forest plant life in urban Singapore. Singapore Univ Press, Kent Ridge, Singapore

Cho D, Boerner REJ (1991) Structure, dynamics, and composition of Sears Woods and Carmean Woods State Nature Preserves, north-central Ohio. Castanea 56:77–89

Clements FE (1916) Plant succession: an analysis of the development of vegetation. Publ 242, Carnegie Inst Wash, Wash

Cole L, Mullard J (1982) Woodlands in urban areas-a resource and refuge. Arboric J 6:295–300

Dawson JO, Khawaja MA (1985) Change in street-tree composition of two Urbana, Illinois neighborhoods after fifty years: 1932–1982. J Arboric 11:344–348

Dean J (2005) "Said tree is a veritable nuisance": Ottawa's street trees 1869–1939. Urb His Rev 24:45–57

Dehnen-Schmutz K, Touza J, Perrings C, Williamson M (2007) The horticultural trade and ornamental plant invasions in Britain. Conserv Biol doi:10.1111/j.1523-1739.2006.00538.x

Denne MP (1987) The tree resource of churchyards of Gwynedd, Wales. Arboric J 11:33–52

Dorney JR, Guntenspergen GR, Keough JR, Stearns F (1984) Composition and structure of an urban woody plant community. Urb Ecol doi:10.1016/0304-4009(84)90007-X

Dougall M, Dickson J (1997) Old managed oaks in the Glasgow area. In: Smout TC (ed) Scottish woodland history. Scottish Cultural Press, Edinburgh, UK

Dunlap TR (1980) The gypsy moth: a study in science and public policy. For His 24:116–126

Esler AE (1991) Changes in the native plant cover of urban Auckland, New Zealand. New Zealand J Bot 29:177–196

Florgård C (2000) Long-term changes in indigenous vegetation preserved in urban areas. Landsc Urb Plan doi:10.1016/S0169-2046(00)00126-2

Forrest M, Konijnendijk CC (2005) A history of urban forests and trees in Europe. In: Konijnendijk CC, Nilsson K, Randrup TB, Schipperijn J (eds) Urban forests and trees: a reference book. Springer, Berlin, Ger

Frawley J (2010) Detouring to Grafton: the Sydney Botanic Gardens and the making of an Australian urban aesthetic. Aust Humanit Rev http://www.australianhumanitiesreview.org/archive/Issue-November-2010/frawley.html. Accessed 12 March 2011

Fry JT (1996) An international catalogue of North American trees and shrubs: the Bartram broadside, 1783. J Gard Hist 16:3–22

Gleason HA (1939) The individualistic concept of the plant association. Am Midl Nat 21:92–110

Grobler CH, Bredenkamp GJ, Brown LR (2002) Natural woodland vegetation and plant species richness of the urban open spaces in Gauteng, South Africa. Koedoe 45:19–34

Grossoni P, Bussotti F, Cenni, E (1999) Tree evaluation in the historic center of a city. Acta Hort 496:45–54

Guldin JM, Smith JR, Thompson L (1990) Stand structure of an old-growth upland hardwood forest in Overton Park, Memphis, Tennessee. In: Mitchell RS, Sheviak CJ, Leopold DJ (eds) Ecosystem management: rare species and significant habitats. Bull 471, New York State Mus, Albany, NY

Haigh H (1984) Pietermaritzburg: the garden city of Natal. Arboric J 8:151–156

Hanson MW (1983) Lords Bushes the history and ecology of an Epping Forest woodland. Essex Nat 7, Plaistrow Press Magazines, London, UK

Hanson T, Davidson E, Mead M, Noonan P (2005) Seward Park vegetation management plan final. Seattle Parks and Recreation, Seattle, Wa. https://www.seattle.gov/parks/Horticulture/VMP/SewardPark.htm. Accessed 7 April 2011

Heineke TE (2009) Floristic study of the Overton Park forest, Memphis, Shelby County, Tennessee. http://spiny.com/naomi/CPOP/OldForestFinalReport.PDF. Accessed 8 March 2011

Hepting GH (1974) Death of the American chestnut. J For His 18:60–67

Hepting GH (1977) The threatened elms: a perspective on tree disease control. J For His 21:90–97

Hoglund AW (1962) Forest conservation and stove inventors: 1789–1850. For His 5(4):2–8

Hytteborn H, Packham JR (1985) Left to nature: forest structure and regeneration in Fiby Urskog, Central Sweden. Arboric J 9:1–11

Jarvis PJ (1973) North American plants and horticultural innovation in England, 1550–1700. Geogr Rev 63:477–499

Jim CY (1986) Street trees in high-density urban Hong Kong. J Arboric 12:257–263

Jim CY (1992) Provenance of amenity-tree species in Hong Kong. Arboric J 16:11–23

Jim CY (2002) Heterogeneity and differentiation of the tree flora in three major land uses in Guangzhou City, China. Ann For Sci doi:10.1051/forest: 2001010

Jim CY, Liu HT (2001) Species diversity of three major urban forest types in Guangzhou City, China. For Ecol Manage doi:10.1016/S0378-1127(00)00449-7

Jones D (1993) Tree planting in the city of London. Q J For 87:37–38, 43–46

Kelly DL (1981) The native forest vegetation of Killarney, south-west Ireland: an ecological account. J Ecol doi:10.2307/2259678

Konijnendijk CC, Ricard RM, Kenney A, Randrup TB (2006) Defining urban forestry – a comparative perspective of North America and Europe. Urb For Urb Green doi:10.1016/j.ufug.2005.11.003

Kunick W (1987) Woody vegetation in settlements. Landsc Urb Plan doi:10.1016/0169-2046(87)90006-5

Lacan I, McBride JR (2009) War and trees: the destruction and replanting of the urban and peri-urban forest of Sarajevo, Bosnia and Herzegovina. Urb For Urb Green doi:10.1016/j.ufug.2009.04.001

Lawrence HW (2006) City trees: a historical geography from the renaissance through the nineteenth century. Univ Virginia Press, Charlotte, Va

Lawrenz KP (1989) Selected arboreta, parks and tree collections in Switzerland. Arboric J 13:321–332

Layton RL (1985) Recreation, management and landscape in Epping Forest: c. 1800–1984. Field Stud 6:269–290

Leighton A (1970) Early American gardens "for meate or medicine". Houghton Mifflin, Boston, Ma

Leighton A (1986) American gardens in the eighteenth century "for use or for delight". Univ Massachusetts Press, Amherst, Ma

Leighton A (1987) American gardens of the nineteenth century "for comfort and affluence". Univ Massachusetts Press, Amherst, Ma

Leland ES, Smith DW (1941) The pioneers of cemetery administration in America. Sterling Press, New York, NY

Leverett R (1996) Definitions and history. In: Davis MD (ed) Eastern old-growth forests: prospects for rediscovery and recovery. Island Press, Washington, DC

Li W, Ouyang Z, Meng X, Wang X (2006) Plant species composition in relation to green cover configuration and function of urban parks in Beijing, China. Ecol Res doi:10.1007/s11284-005-0110-5

Loeb RE (1982) Reliability of the New York City Department of Parks and Recreation's forest records. Bull Torrey Bot Club 117:537–541

Loeb RE (1989) Lake pollen records of the past century. Palynology 13:3–19

Loeb RE (1992a) Long-term human disturbance of an urban park forest, New York City. For Ecol Manag doi:10.1016/0378-1127(92)90142-V

Loeb RE (1992b) Will a tree grow in Brooklyn? Developmental trends of the New York City street tree forest. J For 90(1):20–24

Loeb RE (2001) Fire in the urban forest: long-term effects in old growth stands. Arboric J 25:307–320

Loeb RE (2010) Diversity gained, diversity lost: long-term changes in woody plants in Central Park, New York City and Fairmount Park, Philadelphia. Stud Hist Gard Des Landsc doi:10.1080/14601170903040819

Loeb RE, Germeraad J, Treece T, Wakefield D, Ward S (2010) Effects of one-year versus annual treatment of Amur honeysuckle in forests. Invasive Plant Sci Manag 3:334–339

Mayhew C (2009) A technique to help arboriculturists understand the sequential nature of tree introductions into historical landscapes. Arboric J 32:51–58

McBride JR, Froehlich D (1984) Structure and condition of older stands in parks and open space areas of San Francisco, California. Urb Ecol 8:165–178

McBride JR Jacobs DF (1986) Presettlement forest structure as a factor in urban forest development. Urb Ecol doi:10.1016/0304-4009(86)90003-3

McBride JR, Mossadegh A (2000) Tree-lined canals and the urban forest of Tehran. Arboric J 24:155–173

McCarroll D, Loader NJ (2004) Stable isotopes in tree rings. Quat Sci Rev doi:10.1016/j.quascirev.2003.06.017

McCubbin ME (2009) The Presidio of San Francisco National Park: identifying significance in a living landscape and competing values and preservation of the Presidio National Park's West Pacific Grove. MS Thesis, Pennsylvania State Univ, State College, Pa

McPherson EG, Luttinger N (1998) From nature to nurture: the history of Sacramento's urban forest. J Arboric 24:72–88

Monge-Nájera J, Pérez-Gómez G (2010) Urban vegetation change after a hundred years in a tropical city (San José de Costa Rica). Rev Biol Trop 58:1367–1386

Nagendra H, Gopala D (2010) Street trees in Bangalore: density, diversity, composition and distribution. Urb For Urb Green doi:10.1016/j.ufug.2009.12.005

Natti T (1979) The thirty-eight hurricane. For Notes 138:2–5

Nowak DJ (1994) Understanding the structure of urban forests. J For 92(10):42–46

Nowak DJ, O'Connor PR (eds) (2001) Syracuse urban forest master plan: guiding the city's forest resource into the 21st century. Gen Tech Rep NE-287, US Dep Agric, For Serv, Northeast For Res Stn, Newton Square, Pa

Numata M (1977) The impact of urbanization on vegetation in Japan. In: Miyawaki A, Tuxen R. (eds) Vegetation science and environmental protection. Maruzen Company, Tokyo

Oguz D (2004) Remaining tree species from the indigenous vegetation of Ankara, Turkey. Landsc Urb Plan doi:10.1016/S0169-2046(03)00153-1

Peglar SM, Fritz SC, Birks HJB (1989) Vegetation and land-use history at Diss, Norfolk, U.K. J Ecol 77:203–222

Peter D (2008) Tree succession planning: modelling tree longevity in Tuttangga/Park 17, the Adelaide park lands. Diss, Univ Adelaide. http://digital.library.adelaide.edu.au/dspace/handle/2440/48538. Accessed 12 March 2011

Pinheiro MHO, de Almeida-Neto LC, Monteiro R (2006) Urban areas and isolated remnants of natural habitats: an action proposal for botanical gardens. Biodivers Conserv doi:10.1007/s10531-005-1133-5

Poole AL (1992) The Wellington Botanic Garden, New Zealand. Arboric J 16:303–315

Popoola L, Ajewole O (2001) Public perceptions of urban forest in Ibadan, Nigeria: implications for environmental conservation. Arboric J 25:1–22

Poracsky J, Houck MC (1994) The metropolitan Portland urban natural resource program. In: Platt RH, Rowntree RA, Muick PC (eds) The ecological city: preserving and restoring urban biodiversity. Univ Mass Press, Amherst, Ma

Prescott RTM (1982) The Royal Botanic Garden, Melbourne, A history from 1845–1970. Oxford Univ Press, Oxford, UK

Profous GV (1992) Trees and urban forestry in Beijing, China. J Arboric 18:145–153

Profous GV, Rowntree RA (1993) The structure and management of the urban forest in Prague Czechoslovakia. I. Growing space in metropolitan Prague. Arboric J 17:1–31

Profous GV, Rowntree RA, Loeb RE (1988) The urban forest landscape of Athens, Greece: aspects of structure, planning, and management. Arboric J 12:83–107

Quigley MF (2002) Franklin Park: 150 years of changing design, disturbance, and impact on tree growth. Urb Ecosyst 6:223–235

Rackham O (1978) Archaeology and land-use history. Epping Forest – the natural aspect? Essex Naturalist (New Series) 2:16–57

Rackham O (1980) Ancient woodland its history, vegetation and uses in England. Edward Arnold, London, UK

Rackham O (2003) Ancient woodland its history, vegetation and uses in England, 2nd edn. Castlepoint Press, Colvend, UK

Rodoni B, Kinsella M, Gardner R, Merriman P, Gillings M, Geider K (1999) Detection of *Erwinia amylovora*, the causal agent of fire blight in the Royal Botanic Gardens, Melbourne, Australia. Acta Hort 489:169–170

Rowntree RA (1998) Urban forest ecology: conceptual points of departure. J Arboric 24:62–71

Sanders RA (1981) Diversity in the street trees of Syracuse, New York. Urb Ecol doi:10.1016/0304-4009(81)90019-X

Saunders A (1969) Regent's Park: a study of the development of the area from 1086 to the present day. David & Charles, New York, NY

Savill PS, Perring CM, Kirby KJ, Fisher N (eds) (2010) Wytham Woods: Oxford's ecological laboratory. Oxford Univ Press, Oxford, UK

Schmid JA (1975) Urban vegetation: a review and Chicago case study. Res Pap 161, Dep Geogr, Univ Chicago, Chicago, IL

Seppä H (1997) The long-term development of urban vegetation in Helsinki, Finland: a pollen diagram from Töölönlahti. Veg Hist Archaeobot doi: 10.1007/BF01261957

Shaughnessy GL (1987) A case study of some woody plant introductions to the Cape Town area. In: Macdonald IAW, Kruger F, Ferrar AA (eds) The ecology and management of biological invasions in southern Africa. Oxford Univ Press, Oxford, UK

Shepherd W, Cook W (1988) The Botanic Garden, Wellington: a New Zealand history, 1840–1987. Millwood Press, Wellington, NZ

Showers KB (2010) Prehistory of southern African forestry: from vegetable garden to tree planta-tion. Environ His 16:295–322

Skeggs S (1999) Various botanical and social factors and their effects on an urban woodland in Reading, Berkshire. Arboric J 23:209–231

Solińska-Górnicka B, Symonides E (1990) Effect of a large city on the structure of coenoelements in a natural woodland in Warsaw. Plant Ecol doi:10.1007/BF00044833

Solotaroff W (1912) Shade-trees in town and cities their selection, planting, and care as applied to the art of street decoration; their diseases and remedies; their municipal control and supervi-sion. John Wiley & Sons, New York, NY

Soutar RG, Peterken GF (1989) Regional lists of native trees and shrubs for use in afforestation schemes. Arboric J 13:33–43

Spencer R (1986) Fashions in street planting in Victoria. Aust Landsc 4:304–309

Stalter R, Kincaid D (2008) A 70-year history of arborescent vegetation of Inwood Park, Manhattan, New York, U.S. Arboric Urb For 34:245–251

Stearn WT (1972) From medieval park to modern arboretum: the Arnold Arboretum and its historic background. Arnoldia 32:173–197

Stein AB, Moxley JC (1992) In defense of the nonnative: the case of the eucalyptus. Landsc J 11:35–50

Stewart GH, Meurk CD, Ignatieva ME, Buckley HL, Magueur A, Case BS, Hudson M, Parker M (2009) URban Biotopes of Aotearoa New Zealand (URBANZ) II: floristics, biodiversity and conservation values of urban residential and public woodlands, Christchurch. Urb For Urb Green doi:10.1016/j.ufug.2009.06.004

Sudha P, Ravindranath NH (2000) A study of Bangalore urban forest. Landsc Urb Plan doi:10.1016/S0169-2046(99)00067-5

Sukopp H, Werner P (1982) Nature in cities. Council of Europe, Strasbourg, Fr

Tello N, Tomalak M, Siwecki R, Gáper J, Motta E, Mateo-Sagasta E (1985) Biotic urban growing condi-tions – threats, pests and diseases. In: Konijnendijk C, Nilsson K, Randrup T, Schipperijn J (eds) Urban forests and trees a reference book. Springer, Berlin, Ger, doi:10.1007/3-540-27684-X_13

Thaiutsa B, Puangchit L, Kjelgren R, Arunpraparuta W (2008) Urban green space, street tree and heritage large tree assessment in Bangkok, Thailand. Urb For Urb Green doi:10.1016/j.ufug.2008.03.002

Tillman DA (1978) Wood as an energy source. Academic Press, New York, NY

Ward JS, Parker GR, Ferrandino FJ (1996) Long-term spatial dynamics in an old-growth deciduous forest. For Ecol Manag 83:189–202

Watts D (1963) Plant introduction and landscape change in Barbados, 1625–1836. Diss, McGill Univ, Montreal, Quebec

Webb R (1998a) Urban forestry in Singapore. Arboric J 22:271–286

Webb R (1998b) Urban forestry in Kuala Lumpur, Malaysia. Arboric J 22:287–296

Welch JM (1994) Street and park trees of Boston: a comparison of urban forest structure. Landsc Urb Plan doi:10.1016/0169-2046(94)90023-X

Whaley JR (1985) Killarney National Park. Arboric J 9:25–31

Whitehead MJ (1982) Trees of Soviet central Asia. J Arboric 8:40–44

Williams M (1966) The parkland towns of Australia and New Zealand. Geogr Rev 56:67–89

Worpole K (2003) Last landscapes: the architecture of the cemetery in the west. Reaktion Books, London, UK

Yau DP (1982) Street trees of Melbourne. Arboric J 6:95–105

Young T (2004) Building San Francisco's Parks, 1850–1930. Johns Hopkins Univ Press, Baltimore, Md

Zipperer WC (2002) Species composition and structure of regenerated and remnant forest patches within an urban landscape. Urb Ecosys doi:10.1023/B:UECO.0000004827.12561.d4

Zipperer WC, Guntenspergen GR (2009) Vegetation composition and structure of forest patches along urban-rural gradients. In: McDonnell MJ, Hahs AK, Breuste JH (eds) Ecology of cities and towns: a comparative approach. Cambridge Univ Press, Cambridge, UK

Zipperer WC, Zipperer CE (2002) Vegetation responses to changes in design and management of an urban park. Landsc Urb Plan doi:10.1016/0169-2046(92)90002-H

Chapter 2
Historical Continuity

Abstract The historical continuity of old growth urban forests in the New York, New York, and Philadelphia, Pennsylvania, metropolitan areas was analyzed to reveal the strengths and weaknesses of historical ecology methods for forest dynamics research. The use of paleopalynological techniques to determine forest changes in the time span of the past century to recent millennia requires thoughtful selection of the sampling site but can reveal unrecorded species losses and introductions as well as regional forest changes related to urbanization. Pre-forest clearance witness tree records provide the basis for comparison to ascertain how the modern old growth urban forest differs from the forest primeval. Historical floras from the period immediately following the forest resetting event are catalogs of arboreal species present at the time of old growth urban forest formation that can be used to determine subsequent species diversity changes. Forest remeasurement studies are limited because methodological uniformitarianism and actualism are difficult to achieve due to insufficient detail in recording prior field procedures. However, comparisons of past data and resampling results provide invaluable information on changes in the canopy, subcanopy, and sapling layers that is essential to understanding forest dynamics and historical continuity.

Keywords Urban forest dynamics • Historical forest ecology • Paleopalynology • Witness trees • Historical flora • Forest resampling • Oak-chestnut forest region • New York City • Philadelphia

Introduction

Research methods in historical ecology have been described for Europe (Agnoletti and Anderson 2000) and the United States (Egan and Howell 2001). Of special note is the intensive focus on historical ecology methodology for forests in

Britain (Rotherham et al. 2008). The primary sources for historical ecology research (written documents, paintings, maps, plans, lithographs, historical pictures, aerial photographs, plaques, monuments, and historical events) often have limited value because little is known concerning the perspective and purpose in creating the record (Peter 2008). Providing a land use history is not the goal of historical forest ecology research. Instead, the goal for research on old growth urban forest is to determine forest dynamics. The research objective when exploring historical records is to acquire data on the species and diameters of trees in the canopy, subcanopy, and sapling layers at different points in time. Also, information on tree, sapling and seedling losses to disturbances, diseases and pests, as well as management activities that removed or added trees, saplings, and seedlings is important.

As noted in the first chapter, a comparative analysis of old growth urban forest dynamics across forest types in a metropolis is simplified in regard to climate factors when all of the study sites are in a single forest region. An advantage can be gained when one metropolis is located in proximity to a second metropolis so that comparisons within the forest region are possible. For example, New York, New York, and Philadelphia, Pennsylvania, are separated by 138 km and are in the oak-chestnut region, but New York is in the glaciated section and Philadelphia is in the piedmont section of the region (Braun 1950). From 1600 to 1900, New York and Philadelphia were the two dominant cities in the eastern United States. Regional historical records indicate woodcutting by the end of the American Revolution caused the loss of all forests from the lands that would become the metropolitan areas of New York and Philadelphia (Hoglund 1962). However, two forests are reported to have survived the forest resetting event: William L. Hutcheson Memorial Forest, Rutgers University, New Jersey (Buell et al. 1954) and the Hemlock Forest, New York Botanical Gardens, New York (Rudnicky and McDonnell 1989).

In the oak-chestnut region, the practice of arboriculture with native species of the eastern United States began soon after 1600 and was dominant to the 1850s (Loeb 2010). The presence of a particular native species in a specific old growth urban forest can be related to the history of arboriculture at the locale (DeCandido and Lamont 2004; Fitzgerald and Loeb 2008). Old growth urban forests across the oak-chestnut region have experienced tree species losses caused by diseases introduced from arboricultural planting such as the chestnut blight (*Cryphonectria parasitica* (Murrill) Barr) in 1893 (Pennsylvania Chestnut Tree Blight Commission 1912) and Dutch elm disease (*Ophiostoma ulmi* (Buisman) Nannf.) in 1936 (Hepting 1977).

Historical forest ecology methods are examined in this chapter using examples from the New York City and Philadelphia metropolitan areas. Forest species composition information at the regional level is revealed by the first three methods presented: paleopalynology, early floras, and witness trees. Next, resampling to obtain forest dynamics information on individual old growth urban forests is assessed. As part of the assessment process, a research review and original research from the oak-chestnut region is given for each of the old growth urban forest types. The closing section of this chapter is an integrative analysis of old growth urban forest dynamics in the Philadelphia and New York City metropolitan areas.

Regional Forest Species Composition

Paleopalynology

The science of paleopalynology examines fossil pollen preserved in peats, lake sediments, and soils. Fossil pollen records have been used in archeology (Dimbleby 1985) and vegetation history spanning the time of human occupation (Behre 1986; Birks et al. 1988; Vera 2000) but not woodland management (Rackham 2003) with one exception (Loeb 1998). A few pollen studies have been performed to analyze the history of regional forest changes during the development of a metropolis (Baker et al. 1978; Loeb 1989a, 1992b, 1998; Peglar et al. 1989; Brande et al. 1990; Seppä 1997).

Undertaking a paleopalynological study without extensive formal training is not a course of action to be pursued. However, learning how to conduct paleopalynological research is not required for the interpretation of pollen diagrams, which is the important component for historical forest ecology research. As noted above, few paleopalynological studies have been conducted in urban settings; therefore, if old growth urban forests are to be studied in a metropolis, then a search for pollen record sites is needed. Once pollen collection and preservation sites have been identified, research into whether the sites have been disturbed (i.e., removal or mixing of the peat, lake sediments, or soil) is essential. Before contacting a paleopalynologist to request an analysis of an undisturbed record, the need for obtaining the most recent portion of the pollen record must be assessed. Traditional core retrieval devices are not effective in sampling the unconsolidated recent sediments representing the modern day. The alternative is a freezing corer designed to control penetration into the sediment so as to retain the entire sample up to and including the sediment–water interface, which is the most recently deposited portion of the record (Loeb 1989a).

Interpretation of pollen diagrams requires some knowledge about pollen deposition. The amount of pollen produced by trees varies by orders of magnitude depending upon the mode of pollination with highest to lowest production being pollination by wind, insects, and self-pollination. Wind-pollinated trees are represented better in pollen diagrams than insect or self-pollinated trees, but tree location closer to the pollen deposition site improves the chances for pollen presence in the core. Increasing areal size of lakes and bogs is related to larger source area for pollen. Organic-rich soils collect pollen from the immediate vicinity. Pollen is transported to a deposition site by the wind and runoff including rainfall intercepted by trees. Although the chemical forming the pollen wall, sporopollenin, usually permits pollen to be well preserved, some genera have thin walls and break apart relatively easily. Differential preservation of pollen can lead to biased pollen samples, which causes paleopalynologists to discount samples with extensive amounts of corroded pollen. One assumption of pollen record interpretation is all of the pollen grains in a sample are deposited in the same time period but vertical transport of pollen occurs in soils, peats, and lake sediments (Traverse 2007).

The results of a pollen analysis are the counts of the pollen types representing the arboreal and non-arboreal taxa. Samples are displayed in three types of pollen diagrams: percentage, concentration, and influx. The oldest sample is found at the bottom of the diagram and the youngest at the top. Percentage diagrams typically exclude aquatic pollen and fern spores from the total pollen count that makes up the 100% for each sample in the diagram. Concentration diagrams display the number of pollen grains or fern spores per cm^3 in order to show the amount of pollen and spores present in each sample. Pollen concentration can be lower for the uppermost sample of the water–sediment interface than samples from the consolidated sediments further down the core. When sufficient radiometric or historical dates are available for a core, pollen influx diagrams are created by adjusting the pollen concentration for the length of time identified for segments of the pollen record. Lead 210 is a useful radiometric method for recent diagrams since the isotope has a half-life of 22.7 years and near-zero radioactivity occurs in seven half-lives. Total lead is valuable for dating records in urban settings because the addition of tetraethyl lead to gasoline occurred from 1932 to 1973 (Nriagu 1978). The loss of a species from the forest, such as chestnut (*Castanea dentata*; plant nomenclature follows Gleason and Cronquist 1991) to the chestnut blight, is observed in a record by a drop in the pollen of the species. The drop from one sample to the next can be assigned a date based on what is known about the species loss in the area of the lake or bog. Finally, the uppermost sample is assigned the year before the record was sampled.

The goal of interpreting pollen diagrams for historical ecology research is to relate transitions in the pollen percentage, concentration or influx to environmental events, or human-caused changes. When several diagrams from a geographic area are available, comparisons are made for a regional synthesis. A note of caution in utilizing results from palynological research: palynologists interpret a pollen diagram based on a single pollen sample from each segment of the record without consideration of the variability that would be revealed by multiple pollen samples of a segment. In Holocene palynology, which includes the historical period, small pollen changes from one segment of a record to the next are often given great interpretative weight. Also, complex statistical analyses are based on the sequence of changes in the pollen counts of a record using the assumption that samples represent equal time periods within a span bounded by two identified dates. For recent (approximately a century) pollen records from the New Jersey–New York region, an assessment of variability in three adjacent 1 cm samples revealed a broad range for pollen percentages and concentrations (Loeb 1989a). Variability among pollen samples lends credence to the cautionary note of Traverse (2007) to not take "the very elegant equations and graphs presented in this area of palynology too literally."

With Traverse's words to the wise in mind, an interpretation of a recent pollen diagram from Lake Surprise, Watchung Reservation, Union County, New Jersey (40° 41′ 0″ N, 74° 23′ 0″W; Loeb 1989a), is given to introduce how paleopalynological

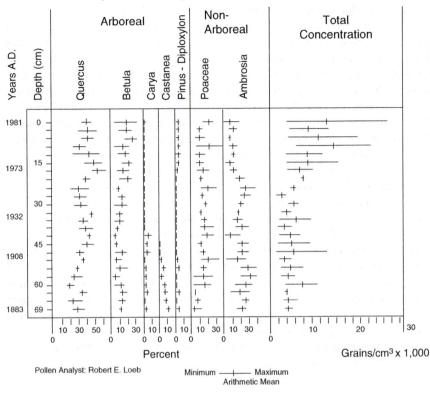

Fig. 2.1 Total pollen concentration and major taxa (greater than 3% mean for any group of three samples) percent pollen diagram for Lake Surprise, Watchung Reservation, Union County, New Jersey (Permission of the American Association of Stratigraphic Palynologists)

analysis applies to old growth urban forest research. Percentage data for the major (greater than 3% mean in any group of three samples) arboreal and non-arboreal genera as well as total concentration data are integrated into one diagram for Lake Surprise (Fig. 2.1). A 72 cm section of sandy clay was obtained from Lake Surprise in early January 1982 using a freezing corer. The top of the lake sediment section is designated as 1981 in the diagram. Sequences of three 1 cm samples were combined to obtain the mean and range with the uppermost segment being used to designate the interval on the diagram (i.e., 69–72 cm was designated as 69). Lead 210 dating provided the date 1883 at the bottom of the record. The chestnut pollen drop between segments 48–51 and 51–54 cm is related to the chestnut blight and dated as 1908 (Pennsylvania Chestnut Tree Blight Commission 1912), which is a revision from the prior designation as 1912 (Loeb 1989a). Addition of tetraethyl lead to gasoline in 1932 corresponds to the increase in total lead between segments 33–36 and

36–39 cm. Removal of tetraethyl lead from gasoline in 1973 is associated with the decrease in total lead between segments 15–18 and 18–21 cm.

In the New York–Philadelphia region, abandoned agricultural fields are first colonized by ragweed (*Ambrosia* spp.), then grass (*Poaceae* spp.), and later trees (Bard 1952). Farm losses to urban expansion is seen as an *Ambrosia* pollen rise in the 66–69 segment followed by a rise in *Poaceae* pollen in the 60–63 cm segment. However, lower percentages for both *Ambrosia* and *Poaceae* pollen begin in the 21–24 cm segment when the expectation is the ragweed decline will precede the grass decrease if rural forests were forming in the region. Instead of rural forests, structures and roads and construction sites were built but did not remain open long enough for ragweed or pollen-producing grass (not lawns) to become established. The *Castanea* pollen decline occurs between 48–51 and 51–54 cm. A *Carya* pollen increase can be observed in the three segments from 42–45 to 48–51 cm, but starting with 39–42 cm *Carya* pollen returns to the lower level before the peak. The canopy opened with the loss of chestnut to the blight and pollen from hickory trees, which do not disperse pollen far (McCarthy and Quinn 1990), was able to enter Lake Surprise without being screened out by chestnut trees. The hickory bark beetle (*Scolytus quadrispinosus* (Say) Solomon) outbreak on Long Island, New York, in 1915 (Anonymous 1916), spread across the region to decimate hickory which is represented by the decrease in *Carya* pollen from 42–45 to 0–3 cm. From segment 21–24 cm to the most recent segment, the higher means for total concentration (primarily arboreal pollen) and pollen percentages for *Betula*, *Pinus* – diploxylon, and *Quercus* represent arboricultural plantings in the growing suburbs of New York City (Solomon and Kroener 1971).

Historical Floras

To understand how the species composition of a regional forest changed after the forest resetting event requires a baseline for comparison, which is a tree species list dated soon after the time of the event. The first urban floras for Philadelphia (Barton 1818) and New York (Torrey 1819) were published 35 years after the end of the American Revolution. Both floras encompassed more than just the cities but the separation between the coverage areas is larger than 50 km. Later floras for the two cities are not appropriate to utilize for comparison because the areal overlap is greater than 50 km (Philadelphia – Keller and Brown 1905; New York – Taylor 1915). The comparison between the two floras is presented in Table 2.1 with nomenclature and designation as alien or native species following Gleason and Cronquist (1991). Because native species from the oak-chestnut forest region were cultivated and used for landscaping (Loeb 2010), Table 2.1 includes the date of first cultivation reported by Rehder (1940). The various terms used to describe the occurrence of each species were reduced to two words "common" and "uncommon" to enable analysis.

The tree species total for the two metropolitan areas was 98. New York had slightly more species (84) than Philadelphia (76). Only 54% of the species were

Table 2.1 Comparison of earliest arboreal floras after forest resetting event for New York, NY (Torrey 1819) and Philadelphia, Pa (Barton 1818)

Species	Cultivation date	New York 1819	Philadelphia 1818
Acer negundo	1688	C	C
Acer pensylvanicum	1755	C	C
Acer rubrum	1650	C	C
Acer saccharum	1753	C	
Alnus rugosa	1769	C	C
Asimina triloba	1736		C
Betula lenta	1759	C	C
Betula lutea	1800	U	
Betula papyrifera	1750	U	C
Betula populifolia	1750	C	C
Carpinus caroliniana	1812	C	C
Carya cordiformis	1650	C	C
Carya glabra	1750	C	C
Carya laciniosa	1800		C
Carya ovata	1629	C	C
Carya tomentosa	1766	C	C
Castanea dentata	1800	C	C
Catalpa bignonioides	1726	C	C
Celtis occidentalis	1636	C	C
Cercis canadensis	1641	C	
Chamaecyparis thyoides	1736	U	
Cornus florida	1730	C	C
Crataegus crus-galli	1691	C	C
Crataegus monogyna (A)	LC	C	C
Crataegus pedicellata	1902	C	U
Crataegus phaenopyrum	1738	C	
Crataegus spathulata	1806		U
Diospyros virginiana	1629		C
Fagus grandifolia	1800	C	C
Fraxinus americana	1724	C	C
Fraxinus carolinana	1724	C	
Fraxinus nigra	1800	C	C
Fraxinus pennsylvanica	1783	C	C
Gleditsia triacanthos	1700	U	C
Hamamelis virginiana	1736	C	C
Ilex opaca	1744	U	U
Juglans cinerea	1633	C	C
Juglans nigra	1686	C	C
Juniperus communis	1560	U	
Juniperus virginiana	B 1664	C	
Liquidambar styraciflua	1681	C	U
Liriodendron tulipifera	1688	C	C
Magnolia virginiana	1688	C	C
Malus coronaria	1724	C	C
Morus rubra	1629	C	

(continued)

Table 2.1 (continued)

Species	Cultivation date	New York 1819	Philadelphia 1818
Myrica cerifera	1699	C	
Nyssa aquatica	B 1735	U	U
Nyssa sylvatica	B 1750	C	C
Ostrya virginiana	1690	C	
Picea mariana	1700	U	U
Pinus rigida	1713	C	C
Pinus strobus	B 1553	U	
Pinus virginiana	B 1739	C	C
Platanus occidentalis	1640	C	C
Populus balsamifera	B 1700	U	
Populus deltoides	B 1750	U	
Populus grandidentata	1772	U	U
Populus heterophylla	1765	U	
Populus tremuloides	1812	U	
Prunus pensylvanica	1773	C	C
Prunus serotina	1629	C	C
Prunus virginiana	1724	C	C
Quercus alba	1724	C	C
Quercus bicolor	1800		C
Quercus coccinea	1691		U
Quercus falcata	B 1763	C	C
Quercus ilicifolia	1800		U
Quercus imbricaria	1786		U
Quercus marilandica	B 1739		C
Quercus muehlenbergii	1822		C
Quercus nigra	1723		U
Quercus palustris	B 1770	C	C
Quercus phellos	1723	U	C
Quercus prinus	B 1800	C	C
Quercus rubra	1691	C	C
Quercus stellata	1819	C	C
Quercus velutina	1800	C	C
Quercus x heterophylla	1882		U
Rhus copallina	1688	C	
Rhus glabra	1863	C	C
Rhus typhina	1898	C	C
Rhus vernix	1713	C	C
Robinia pseudoacacia	1635	C	C
Salix alba (A)	LC	C	C
Salix babylonica (A)	1730	C	
Salix discolor	1809		C
Salix elaeagnos (A)	LC	C	
Salix eriocephala	1898	C	C
Salix lucida	1830	C	
Salix nigra	1809	C	
Salix petiolaris	1802		C

(continued)

Table 2.1 (continued)

Species	Cultivation date	New York 1819	Philadelphia 1818
Salix viminalis (A)	LC		C
Sassafras albidum	1633	C	C
Thuja occidentalis	1536	C	
Tilia americana	1752	C	C
Tsuga canadensis	1736	U	C
Ulmus americana	1752	C	C
Ulmus rubra	B 1830	U	U

Nomenclature and identification of a species as alien (A) follows Gleason and Cronquist (1991). Date of first cultivation is from Rehder (1940) with the code LC meaning long cultivated (at least prior to 1500) and the letter B preceding a date signifies cultivation began before the year indicated. Occurrence of the species is common (C) or uncommon (U)

reported in both floras which indicates the arboreal floras of the two metropolitan areas were quite different. Considering the five alien species, two were found in both floras and two were present only in the Philadelphia flora. Nearly half of the species (48) had the occurrence term common in both floras, while only five species were uncommon in the two floras. Among the species in only one flora, 21 were common and 15 were uncommon. Of the nine species with different assessments of occurrence, five had common for Philadelphia and uncommon for New York. Just nine species had a first cultivation date after 1818, indicating many of the native species that were reported to have grown spontaneously in the two metropolitan areas could have originated in arboricultural plantings at particular sites.

Witness Tree Records

Witness trees recorded in federal land survey documents from Canada and the United States have been used to determine forest composition for the time period just before the European settlement (Miller 1965). The trees are noted as boundary line or corner markers in the federal land survey records which do not include the New York and Philadelphia metropolitan areas. Pre-European settlement (in the locale of the survey) witness tree records in the oak-chestnut forest region (Collins 1956; Greller 1972; Russell 1979; Loeb 1987) may have limited value because of possible surveyor bias in the selection of trees and misidentification of tree species. To ascertain if witness tree records in the New York and Philadelphia metropolitan areas can be used to differentiate between forest regions, the county and township summaries with more than 100 witness trees (Loeb 1987) are compared for the oak-chestnut region and adjacent pine-oak region (Braun 1950). Witness tree records are available for two areas in the oak-chestnut region, southeast New York (4 groups of records) and north East New Jersey (14 groups of records), while only south East New Jersey (5 groups of records) exists to represent the oak-pine region. The two-tailed T-test of independent group means was used to determine if significant differences

Table 2.2 Major taxa (>2% of at least one survey group) means, standard deviations (St Dev), number of surveys (*N*), and significant results at 0.05 level for two-tailed T-test between means of witness tree surveys from 1665 to 1790 in counties and townships of southeast New York (SENY), north East New Jersey (NENJ), and south East New Jersey (SENJ; data from Loeb 1987)

	SENY	*N*=4	NENJ	*N*=14	SENJ	*N*=5
	Mean	St Dev	Mean	St Dev	Mean	St Dev
Acer spp.	1.3	1.3	3.6	2.3	5.0	4.1
Betula spp.	0.5	0.5	2.2	2.1	0.4	0.5
Carya spp.	0.8	1.3	5.8	5.3	1.6	2.2
Castanea dentata	10.8	7.3	6.6	3.5	2.4*	2.1
Chamaecyparis spp. and *Juniperus* spp.	0.3	0.4	0.1	0.3	6.2**	5.7
Fraxinus spp.	2.3	2.8	2.1	1.5	0.6	0.5
Juglans spp.	5.5	4.7	3.9	4.4	1.0	1.1
Pinus spp.	0.0	0.0	0.9	0.5	26.2**	16.6
Quercus spp.	67.0	9.0	63.6	12.6	47.0***	18.1
Quercus rubra	13.5****	5.9	7.7	4.2	6.4	2.9
Quercus velutina	17.3	5.4	14.6	3.2	9.0*	3.4
Quercus alba	33.3	5.4	39.6	9.6	29.0	14.5

*Mean for SENJ is significantly less (at 0.05 level) than means for SENY and NENJ
**Mean for SENJ is significantly greater (at 0.05 level) than means for SENY and NENJ
***Mean for SENJ is significantly less (at 0.05 level) than mean for NENJ
****Mean for SENY is significantly greater (at 0.05 level) than means for NENJ and SENJ

exist between the groups of records for the major (>2% of any group) tree taxa (surveyor use of common names limits association with species binomials). The *T*-tests calculations were performed using PASW Statistics (formerly SPSS Statistics) version 17 and the significance level selected was 0.05.

Examining the witness tree records (Table 2.2) for significant differences in the percentages of the codominants chestnut and pine (*Pinus* spp.) can serve to differentiate the two regions (oak-chestnut versus oak-pine). The means for chestnut in north East New Jersey and southeast New York are significantly greater than the mean for south East New Jersey. For pine, the south East New Jersey mean is significantly larger than the means for north East New Jersey and southeast New York. Cedar (*Chamaecyparis* spp. and *Juniperus* spp.) is the taxon in the witness tree records associated with the Atlantic white cedar (*Chamaecyparis thyoides*) swamp forests of the oak-pine region (Braun 1950). The mean for cedar in the south East New Jersey group is significantly greater than the means for north East New Jersey and southeast New York. For black oak (*Quercus velutina*), the means in both oak-chestnut region areas are significantly larger than the mean for the oak-pine region groups, which appears to be related to the preference of black oak for the moist-rich soils of the oak-chestnut region over the sandy dry soils typical of the oak-pine region (Hannah 1968). Northern red oak (*Quercus* rubra) had a significantly larger mean in the southeast New York survey group than both survey groups for East New Jersey. Perhaps Native Americans caused greater tree mortality in southeast New York than north and south East New Jersey because northern red oak responds well to canopy openings (Graney 1987). In summary, the colonial period witness tree records showed significant differences among the major taxa that indicate the oak-chestnut and the oak-pine regions existed prior to European settlement (Table 2.2).

Forest Dynamics

Analysis of forest dynamics requires comparisons to determine the past, present, and future of the forests in a metropolitan area. The goal is to identify several sites with differing ecological histories for each old growth urban forest type. Although all sites should be considered and forest measurements begun where none existed before, old growth urban forests with previous forest research present the opportunity to compare the current forest with past conditions (Peterken and Backmeroff 1988). Two principles apply when prior studies are available for comparison to a remeasurement of the forest: methodological actualism and uniformitarianism. Methodological actualism refers to performing the forest remeasurement with the methods utilized in the prior study. Duplication of methods and relocation of sample sites is needed for comparisons of long-term changes because two types of forest variability exist. Henry Gleason's (1939) description of the two types is "In time-variability, the environment changes from one time to another on the same spot. In space-variability, the environment differs from spot to spot at the same moment." For example, in 1935, three different results were produced by using three different sampling methods in the St John's Pond Forest, Long Island, New York, and when resampling was conducted in 1982, the forest changes through time were also different (Loeb 1990). Methodological uniformitarianism refers to using the same methods at all sites being compared, which is difficult at best for historical ecology research considering the variety of forest measurement procedures used in the past. Even if methodological uniformitarianism cannot be achieved with field methods, comparisons of results must be made on the same type of statistics, such as stems per hectare or percentage of stems.

The ideal methodology is annual forest inventory but considering the lifespan of trees and the expense of conducting an inventory, continuous forest inventory is sufficient as long as the periodic measurements are performed as scheduled. Repeated measures statistical methods (Crowder and Hand 1990) can be applied to the data as well as analysis of changes in the forest distribution (Cho and Boerner 1991; Ward et al. 1996). The forest inventory records the location, size, and condition of every member of the canopy, subcanopy, and sapling classes within the entire area designated as the forest (McBride and Nowak 1989). Seedlings are counted in sections of the forests but not located until the individual stems join the sapling class. When a high-resolution global positioning system (GPS) and a large capacity geographic information system (GIS) are available, precise measurements of tree dimensions and position can be recorded, which permits concerns about forest variability introducing uncertainty into measurements of changes through time to be safely set aside. More relevant to historical forest ecology research is how recently and infrequently GPS and GIS technology has been used. Past forest records do not have precise location information or the capacity to correct field investigator errors. For example, the New York City Department of Parks and Recreation has a park map series created in the 1930s which gives the location, species identification, and diameter measure for all of the trees. A remapping of Seton Falls Park, Bronx, New York, using the 1936 map revealed inaccurate locations, misidentifications to the family level, and diameter measures indicating incredible rates of tree growth and shrinking trees (Loeb 1982). Forest dynamics research employing historical forest maps must account for errors, and comparative analysis is limited to reliable information (Stalter and Kincaid 2008).

Historical records of urban forest composition rarely involve maps locating trees. Instead the common record used for comparison in resampling research has an unclear description of method and imprecise location of sampling points. Often there is nothing more than a total for each species and a location associated with the information which could be an ill-defined tract for remnant and landscaped forests or entire cities for street forests. The following historical forest ecology research studies and literature reviews for the three old growth urban forest types show forest dynamics based on resampling; however, true methodological actualism and uniformitarianism was not possible because of the methods' descriptions, or lack thereof, in the prior research reports.

Street Forest

Original Research: East Orange, Haddonfield, and Moorestown, New Jersey

Solotaroff (1912) described the state of the art for urban shade-trees planting and maintenance in the New York and Philadelphia metropolitan areas at the beginning of the twentieth century. Also, Solotaroff provided a summary of the East Orange, New Jersey, street forest species (Table 2.3) and reported the dbh (diameter at breast height, 130 cm) distribution as 1,698 trees under 15.2 cm dbh, 7,036 trees from 15.2 to 45.7 cm dbh, and 2,219 trees greater than dbh 45.7 cm. In 2004, a survey of the entire street forest was performed but no dbh data were recorded (Anonymous 2006). From 1911 to 2004, the total number of stems dropped by 40%. The number of species present in 1911 but not reported in 2004 was 24, and 10 species were noted in 2004 but not in 1911. The ratio of alien to native taxa in 1911 of 17:33 and 10:19 in 2004 demonstrates a decline for both alien and native species diversity. Red maple (*Acer rubrum*), silver-maple (*Acer saccharinum*), and sugar maple (*Acer saccharum*) went from comprising over two-thirds of the forest in 1911 to less than one-tenth in 2004. The stems per hectare (sph) for Norway maple (*Acer platanoides*) declined slightly from 1911 to 2004 but rose from 11% to 17% of the forest. Lindens (*Tilia* spp.), plane trees (*Platanus* spp.), and pin oak (*Quercus palustris*) joined Norway maple to comprise the four most frequent taxa

Table 2.3 Street tree taxa stems per hectare for East Orange, New Jersey in 1911 (Solotaroff 1912) and 2004 (Anonymous 2006) as well as Haddonfield, New Jersey and Moorestown, New Jersey in 2010

	First date cultivated	East Orange Year of Inventory		Haddonfield			Moorestown		
		1911	2004	Sapling	Subcanopy	Canopy	Sapling	Subcanopy	Canopy
Acer negundo	1688	0.013	0	0.014	0.030	0.014	0	0.006	0.001
Acer platanoides (A)	LC	1.185	1.125	0.088	0.508	0.230	0.002	0.066	0.019
Acer pseudoplatanus (A)	LC	0.049	0.014	0	0.010	0.007	0	0	0
Acer rubrum	1650	2.790	0.485	0.127	0.647	0.29	0.033	0.250	0.050
Acer saccharinum	1725	2.183	0.044	0.003	0.015	0.093	0	0.013	0.028
Acer saccharum	1753	2.187	0.068	0.026	0.448	0.655	0.003	0.057	0.058
Aesculus glabra	1809	0.007	0	0.004	0	0	0.003	0.009	0
Aesculus hippocastanum (A)	1576	0.173	0	0.001	0.003	0.007	0.002	0.005	0.001
Ailanthus altissima (A)	1784	0.006	0.001	0.001	0.004	0.005	0	0	0
Betula nigra	1736	0	0.008	0.005	0.004	0.001	0	0.001	0
Carpinus caroliniana	1812	0.004	0	0	0.004	0	0.001	0.023	0.001
Carya spp.		–	0.023	0	0	0	0	0.003	0.002
Carya cordiformis	1650	0	–	0	0.026	0.008	0	0.001	0
Carya glabra	1750	0.006	–	0	0	0.001	0	0	0
Carya ovata	1629	0.002	–	0	0	0.001	0	0	0
Carya tomentosa	1766	0.002		0	0	0	0	0	0
Castanea mollissima (A)	1854	0	0.033	0	0	0	0	0	0
Catalpa spp.	1730	0.028	0	0.001	0.003	0.011	0	0	0
Cornus florida	1800	0.003	0	0.034	0.148	0	0	0.001	0
Fagus grandifolia	1800	0.006	0.001	0.003	0.005	0.008	0	0.009	0.01
Fraxinus americana	1724	0.078	0.159	0.03	0.249	0.229	0.003	0.022	0.004
Ginkgo biloba (A)	1784	0	0.240	0.003	0.004	0.008	0	0.015	0.001
Gleditsia triacanthos	1700	0.001	0.175	0.029	0.066	0.047	0.007	0.032	0.001
Halesia tetraptera	1756	0.001	0	0	0	0	0	0.002	0

(continued)

Table 2.3 (continued)

	First date cultivated	East Orange Year of Inventory		Haddonfield			Moorestown		
		1911	2004	Sapling	Subcanopy	Canopy	Sapling	Subcanopy	Canopy
Juglans cinerea	1633	0.001	0	0.001	0.011	0.022	0	0	0
Juglans nigra	1686	0.002	0	0	0.001	0.012	0	0.002	0.002
Liquidambar styraciflua	1681	0.005	0.020	0.015	0.048	0.386	0.001	0.020	0.011
Liriodendron tulipifera	1688	0.009	0.007	0.022	0.149	0.140	0	0.005	0.004
Morus alba (A)	LC	0.001	0	0	0	0	0	0	0
Nyssa sylvatica	B 1750	0.001	0	0.003	0.01	0.005	0	0.001	0.001
Picea abies (A)	LC	0.002	0	0.001	0.045	0.062	0	0.002	0.001
Pinus strobus	B 1553	0.005	0	0.004	0.089	0.071	0.001	0.031	0.001
Platanus spp.	–	–	0.802	0	0	0	0	0	0
Platanus occidentalis	1640	0.022	–	0.027	0.01	0.455	0.004	0.020	0.041
Platanus orientalis (A)	1842	0.006	–	0	0	0	0	0	0
Platanus x acerifolia (A)	B 1700	0	–	0.011	0.007	0.004	0.005	0.027	0.080
Populus alba (A)	LC	0.007	0	0	0	0	0	0	0
Populus x canadensis (A)	1818	0.713	0	0	0	0	0	0	0
Populus grandidentata	1772	0.001	0	0	0.003	0	0	0	0
Populus nigra (A)	LC	0.002	0	0	0	0.001	0	0	0
Prunus cerasus (A)	LC	0.002	0	0	0	0	0	0	0
Prunus serotina	1629	0.004	0	0.018	0.152	0.048	0.002	0.008	0.003
Prunus serrulata (A)	1900	0	0.063	0.010	0.138	0.086	0	0.002	0
Pyrus calleryana (A)	1908	0	0.179	0.015	0.168	0.068	0.001	0.013	0.005
Pyrus malus (A)	LC	0.001	0	0	0	0	0	0	0
Quercus alba	1724	0.019	0.015	0.001	0.003	0.015	0.001	0.003	0.002
Quercus bicolor	1800	0.011	0	0.037	0.033	0	0	0	0
Quercus macrocarpa	1811	0	0.017	0	0.014	0.005	0	0.001	0.001
Quercus palustris	B 1770	0.022	1.339	0.003	0.145	1.297	0.003	0.047	0.116
Quercus rubra	1691	0.011	0.312	0.012	0.262	0.733	0.001	0.008	0.027

	Date								
Quercus velutina	1800	0.004	0.003	0	0.027	0.021	0.003	0.082	0
Salix babylonica (A)	1730	0	0	0	0	0	0	0	0.004
Sassafras albidum	1633	0.001	0.002	0.001	0.003	0.007	0.003	0	0.002
Sophora japonica (A)	1747	0	0.001	0	0.012	0.011	0.001	0.126	0
Sorbus americana	1811	0	0	0	0	0	0	0.007	0
Thuja occidentalis	1536	0	0.008	0.002	0.011	0.068	0.007	0	0.001
Tilia spp.				0.001	0	0	1.059	—	—
Tilia americana	1752	0.002	0.016	0.007	0.071	0.034	0.003	—	0.130
Tilia cordata (A)	LC	0.037	0.047	0.002	0.118	0.181	0.059	—	0.045
Tilia tomentosa (A)	1767	0	0	0	0.008	0.064	0.003	—	0
Ulmus spp.								0.017	—
Ulmus americana	1752	0.001	0.005	0	0.059	0.086	0.007	—	0.974
Ulmus parvifolia (A)	1794	0.001	0.001	0	0.001	0.015	0.001	—	0
Ulmus procera (A)	B 1770	0	0	0	0.001	0	0	—	0
Ulmus pumila (A)	1860	0	0	0	0.023	0.015	0	—	0
Ulmus rubra	B 1830	0	0.001	0	0	0.004	0	—	0.002
Zelkova serrata (A)	1862	0.001	0	0.006	0.16	0.188	0.047	0.057	0
Other species		0.174	0.461	0.250	0.576	1.112	0.391	0	0
Total Stems Per Hectare		0.692	1.250	0.341	6.096	5.268	1.074	6.475	10.727
Total Trees		1591	2873	784	4450	3849	784	6605	10.953

Native species nomenclature follows Gleason and Cronquist (1991) and alien species nomenclature follows Rehder (1940). The letter A in parenthesis after the species name indicates an alien species. Date of first cultivation is from Rehder (1940) with the code LC meaning long cultivated (at least prior to 1500) and the letter B preceding a date signifies cultivation began before the year indicated. The Haddonfield, New Jersey, and Moorestown, New Jersey, data are in the dbh size classes: sapling (0 cm<dbh<7.5 cm), subcanopy (7.5 cm≤dbh≤45.7 cm), and canopy (dbh>45.7 cm) in 2004. All of the species reported for East Orange, New Jersey, in either 1911 or 2004 entered cultivation before 1908. Similar changes in species composition were observed in the New York City street forest; however, plantings and removals from 1900 to 1990 kept the size of the forest nearly stable throughout the 90-year period (Loeb 1992a).

Tree survey records were obtained for the street old growth urban forests in Haddonfield, New Jersey, and Moorestown, New Jersey (60% completed in 2010). The survey data were converted into stems per hectare and divided into three dbh size classes: sapling (0 cm < dbh < 7.5 cm), subcanopy (7.5 cm ≤ dbh ≤ 45.7 cm), and canopy (dbh > 45.7 cm). To enable comparison with East Orange, the data for the two towns were incorporated into Table 2.3 but no major taxon (>5% of the forest) from the Haddonfield and Moorestown surveys was excluded. The total for taxa in Haddonfield was 144 as compared to 131 for Moorestown and the two towns had 91 species in common. Despite the great species diversity in Haddonfield and Moorestown, eight species from 1911 and two species from 2004 were present in East Orange but not Haddonfield and Moorestown. The species with sph > 0.25 for Haddonfield were white ash (*Fraxinus americana*), callery pear (*Pyrus calleryana*), sycamore (*Platanus occidentalis*), Norway maple, red maple, sugar maple, pin oak, and northern red oak. For Moorestown, only red maple had an sph > 0.25.

Two characteristics are useful for ascertaining if a street forest is an old growth urban forest: for all species the largest size class (dbh>45.7 cm) has the highest sph and the species with the greatest sph match the species in historical forests (East Orange in 1911) or historical floras (Table 2.1). The largest size class is nearly half of the Haddonfield forest but less than a third of the Moorestown forest. For East Orange, the species in 1911 with sph > 0.25 were Norway maple, red maple, silver-maple, sugar maple, Carolina popular (*Populus x canadensis*), and American elm (*Ulmus americana*), while the species in 2004 with sph > 0.25 were Norway maple, red maple, plane trees, pin oak, northern red oak, and lindens. The Haddonfield species with sph > 0.25 match all of the taxa having an sph > 0.25 from East Orange in 1911 except silver-maple, American elm, and Carolina poplar, and in 2004 except lindens. Haddonfield fits the characteristics of a street old growth urban forest. In contrast Moorestown does not because of the low sph in the largest size class and only red maple having an sph > 0.25, which is a common element to both the 1911 and 2004 species with sph > 0.25 in East Orange. The problem with making a judgment concerning Moorestown is the lack of historical perspective on changes in the forest. For example, the populations of pin oak and northern red oak have been decimated by oak wilt (*Xyella fastidiosa*) in Moorestown (but not Haddonfield), and during the past 5 years, 20 different species have been planted and none are species with sph > 0.25 for East Orange (Gibson et al. 2010). Diversifying a street forest is beneficial to lessen the chance of major losses to disease (e.g., Dutch elm disease), but planting trees to maintain the population of viable old growth native species is the essential planning goal to achieve historical continuity.

Landscaped Forest

Original Research: Beechwood, Country Club, and Robert's Hollow Forests, Fairmount Park, Philadelphia, PA

Fairmount Park is located north of center city Philadelphia, Pennsylvania (40° 0′ N; 75° 12′ W), and the Schuylkill River divides the Park into two sections: East and West Fairmount Park. Three landscaped old growth urban forests are present in the northern portion of West Fairmount Park: Beechwood, Country Club, and Robert's Hollow. An 1868 map (Fairmount Park Commission 1868) depicts the forests as does a 1900 map (Fairmount Park Commission 1900), which also includes the former park trolley system (Fig. 2.2). Research for the historical ecology of Fairmount Park was conducted in the archives of the Fairmount Park Commission, City of Philadelphia, and Dickinson College and the libraries of the Academy of Natural Sciences of Philadelphia, Historical Society of Pennsylvania, New York Botanical Gardens, and Pennsylvania Horticultural Society. Because the historical record is exceptionally rich for old growth urban forests, the information will be divided in two sections: historical ecology and forest dynamics.

Historical Ecology

Before 1770, estates were built and landscaped in the land that would become Fairmount Park, and during the American Revolution the land was virtually cleared of trees. The estates were replanted with native species trees (not just saplings and seedlings) prior to 1800 (Loeb 2010) and were incorporated into the Park before 1868 (Anonymous 1868; Fig. 2.3). Although no records of the species planted in the estates are available, the trees commonly planted in the region prior to 1800 (Adams 2004) were as follows: balsam fir (*Abies balsamea*), red maple, sugar maple, red buckeye (*Aesculus pavia*), sweet birch (*Betula lenta*), southern catalpa (*Catalpa bignonioides*), flowering dogwood (*Cornus florida*), eastern red cedar (*Juniperus virginiana*), tulip-tree (*Liriodendron tulipifera*), black spruce (*Picea mariana*), white pine (*Pinus strobus*), white oak (*Quercus alba*), scarlet oak (*Quercus coccinea*), black oak, black locust (*Robinia pseudoacacia*), sassafras (*Sassafras albidum*), and eastern hemlock (*Tsuga canadensis*). The estates should be expected to have had greater native species diversity than others in the oak-chestnut region because Bartram's Garden, the historical origin of native species stock for arboricultural plantings in the United States (Fry 1996), was nearby.

Landscaping in East Fairmount Park from 1860 to 1868 was focused on planting groups of native species into the existing forest of native species (Sidney and Adams 1859). In planning for further landscaping, a survey of the trees revealed 31,421 trees with dbh of 45.7 cm or greater. The total of smaller dbh trees was estimated to be more than 90,000 but no enumeration by species was done (Cresson 1868). Plantings in the 1876 Centennial Fairgrounds focused on alien angiosperm

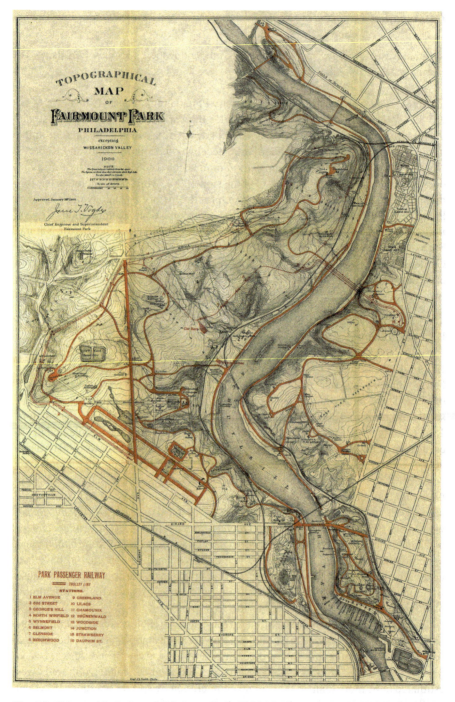

Fig. 2.2 Topographical map of Fairmount Park, Philadelphia, excepting Wissahickon Valley, 1900 (Fairmount Park Commission 1900; permission of the Fairmount Park Historic Resource Archive; Collection of Adam Levine, www.phillyh2o.org)

Fig. 2.3 Map of farms and lots embraced within the limits of Fairmount Park (Anonymous 1868, with permission of Collection of Adam Levine, www.phillyh2o.org)

and gymnosperm species which enabled Fairmount Park to have more tree and shrub species than Central Park, New York City, in 1880. By 1970, Fairmount Park lost more than half of the species present in 1880 while Central Park had more species in 1970 than 1873. From 1970 to 2007, 60 tree and shrub species were lost from the Centennial Fairgrounds (Loeb 2010).

The extensive arboricultural plantings made in preparation for the 1876 Centennial Fair (Rothrock 1880) and later (Corson 1937) did not affect the three landscaped old growth urban forests because of their location away from the Centennial Fairgrounds (Fig. 2.2). Fires started by sparks from railroad engines began with the completion of the Reading Railroad tracks between the west shore of the Schuylkill River and West Fairmount Park on December 9, 1839 (Holton 1989), and ceased with the complete replacement of steam locomotives with diesel locomotives by 1955 (Marx 1976). Sweetbrier (*Rosa eglanteria*) and catbrier (*Smilax rotundifolia*) were so common in the Park that the grass cutting contract for 1895 incorporated specifications for cutting briars (Fairmount Park Commission 1895). Pedestrian access to the Beechwood and Country Club forests was increased by the Fairmount Park Trolley (1897–1946; Cox 1970) having stops at both sites (Fig. 2.2 – note the Country Club forest is at the Brünenwald trolley stop) but the location of Robert's Hollow in the most northern point of the Park was not a trolley stop. The Fairmount Park Stable is located next to Country Club and horseback riders continue to be permitted to use the trails in the forest.

In 1908, Oglesby Paul, landscape gardener of Fairmount Park, reported on the condition of the trees in the Park including the Beechwood, Country Club, and Robert's Hollow forests (Paul 1908). Insect infestations caused deaths for Norway maple, silver-maple, southern catalpa, beech (*Fagus* spp.), walnut (*Juglans* spp.), black gum (*Nyssa sylvatica*), oriental plane-tree (*Platanus orientalis*), white oak, pin-oak, rock chestnut-oak (*Quercus prinus*), black willow (*Salix nigra*), linden, and American elm. In Beechwood, the chestnut blight caused a few trees to be "removed in the year 1907, since when many more have died." The Beechwood forest was reported to be in poor condition because fires burned the humus layer and killed seedlings and saplings; the public trampled the seedlings and saplings; and sprout growth was short lived. Fires started by railroad engines in the winter of 1907 burned half of Beechwood and the entire area of Robert's Hollow. Prior to the fire, the Robert's Hollow forest had a thick layer of humus. Paul wrote "At the Country Club alone do we find a perfectly healthy forest, with an adequate proportion of forest cover, veterans, and saplings to insure its future existence and character." Furthermore, Paul noted "One of the chief reasons for the excellent condition of the Country Club woods is the thorough forestry work we did there three years ago. Numbers of windfalls and dead and dying trees were cut out at that time, thus affording a chance for the surrounding trees to seed into open spaces, and also admitting sunlight and air to the young saplings already struggling for a foothold." Although not reported as carried out, Paul recommended removing competitors of oak (*Quercus* spp.) and walnut in thickets of seedlings.

The purpose of Paul's report was an appeal for an increase in the funding for arboricultural work in Fairmount Park. In 1907, $3,500 was allotted for tree work in Fairmount Park as compared to $50,000 for Central Park, New York, which was a quarter of the size of the area under Fairmount Park Commission control (Paul 1908). Over the decades of the twentieth century, the Fairmount Park Commission became responsible for the street forest and the expanding park system of Philadelphia, and as the years passed, relatively less attention was paid to Fairmount Park. The Fairmount Park Commission funded an ecological study of the Park in 1969 which recommended protection of the Beechwood, Country Club, and Robert's

Hollow forests (McCormick 1971). White-tailed deer (*Odocoileus virginianus*; mammal nomenclature follows Kays and Wilson 2009) returned to the area of Fairmount Park in the 1980s and the Commission initiated a culling program in 1999 (Brown 2005).

As noted above, the 1868 survey included all of the trees with dbh greater than 45.7 cm and was for the entire Fairmont Park (Cresson 1868), but at the time Robert's Hollow was not included in the Park. Based upon the forest inventory, chestnut was the species with the largest stems per hectare (sph) in 1868 at 21.5 for trees with dbh> 45.7 cm in Fairmount Park. Sugar maple and silver-maple, eastern red cedar, black mulberry (*Morus nigra*), and white oak had an sph greater than 13. Of the 17 remaining taxa with an sph greater than two, only bitternut-hickory (*Carya cordiformis*), southern catalpa, and tulip-tree had an sph greater than 6.5 (Fig. 2.4).

Paul's report (1908) had 1907 survey results for 18 forests including Beechwood, Country Club, and Robert's Hollow. The survey methods description was incomplete and the location of sampling sites was not given (Paul 1908), which precluded duplication of methods. The remeasurement method selected was a total survey of the largest minimum area located at least 10 m from the edges of the forest. Since the entire area of Robert's Hollow (1.9 ha) was surveyed in 1907, a 60 m by 300 m plot (1.8 ha) was placed within the old growth urban forest of Robert's Hollow in 2007. In the larger forests at Beechwood and Country Club (in 1907, 10 ha and

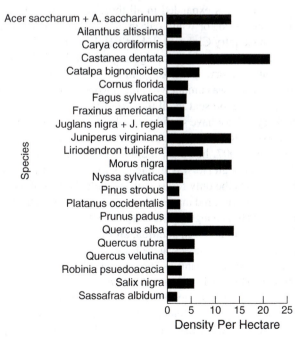

Fig. 2.4 Fairmount Park trees (dbh>45.7 cm) in 1868 (Cresson 1868) with stems per hectare greater than two

12 ha, respectively), a 60 m by 300 m plot was placed within the remaining area
(losses due to buildings and roadway construction) for consistency in survey method
with Robert's Hollow. The 1907 data were published in diameter size classes (Paul
1908). In order to permit comparison with the 1868 data and to examine forest
dynamics, three size classes were selected to report: canopy trees (dbh > 45.7 cm),
subcanopy trees (45.7 cm ≥ dbh ≥ 10.2 cm), and saplings (10.2 cm > dbh > 2.5 cm).
The data from each time period were converted to stems per hectare.

In 1907, chestnut, northern red oak, and black oak in the canopy, and choke-cherry
(*Prunus virginiana*) in the subcanopy (Fig. 2.5a–d, respectively) had their highest
sph in Beechwood, a lower sph in Country Club, and the lowest sph in Robert's
Hollow. Each of the four species had very few or no saplings in 1907. By 2007,
chestnut and choke-cherry were absent while northern red oak and black oak had
more than half of their populations in the largest size class. In 2007, black oak sph
followed the pattern among the forests from 1907. In contrast, the pattern for 1907
and 2007 differed for northern red oak with the total sph for each forest being almost
equal, and saplings were present in all three forests in 2007.

Again in 1907, American beech, tulip-tree, and white oak in the canopy, and
flowering dogwood in the subcanopy (Fig. 2.6a–d, respectively) had their highest
sph in Country Club. Among the four species, only white oak had a lower sph in
Beechwood than Robert's Hollow and was distinctive as being the only species with
no saplings in all three forests. Flowering dogwood virtually disappeared from all
three forests by 2007. American beech decreased in each size class in all three forests except for saplings in Country Club and Robert's Hollow. Tulip-tree's total sph
was much closer to equal among the forests in 2007 than 1907, and the population
of canopy trees expanded in all three forests by 2007. The number of subcanopy
trees increased for white oak in 2007 relative to 1907, and saplings were present
only in Country Club and Robert's Hollow in 2007.

The species with their largest sph in Robert's Hollow were rock chestnut-oak,
bitternut-hickory, white ash, and Norway maple (Fig. 2.7a–d, respectively). Rock
chestnut-oak was uniquely found in Robert's Hollow in 1907 with an sph of 40, and
by 2007 was present in all three forests but had an sph of less than one. Bitternut-
hickory did not have saplings in 1907 but two-thirds of the population was saplings
in 2007. In Robert's Hollow, white ash had only canopy and subcanopy trees in both
1907 and 2007; however, the total sph was nearly 60 times greater in 2007 than
1907. Although Norway maple was found in all three forests just in 2007, Robert's
Hollow was the only forest with individuals in all three size classes.

Four species, red maple, black gum, wild black cherry, and sassafras (Fig. 2.8a–d,
respectively), changed from being absent in at least one forest in 1907 to being found
in all three in 2007. In addition, all four species had a larger sph in 2007 than 1907
for all three size classes in each forest, except for the greater sph for sassafras subcanopy trees in Country Club and wild black cherry had a higher sapling sph for
Robert's Hollow. Hornbeam (*Carpinus caroliniana*) was found only in Country Club
and had a threefold increase in subcanopy and canopy tree sph to approximately 30
but no saplings were found in 1907 or 2007. For the species with a total sph greater
than 10 that were first recorded in 2007, Norway maple, Hercules club (*Aralia*

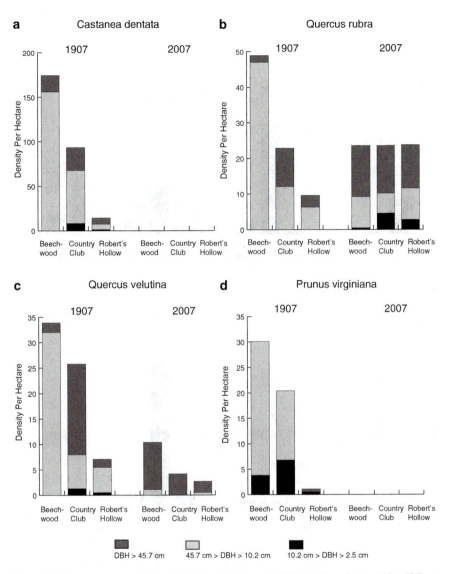

Fig. 2.5 Stems per hectare for 1907 (Paul 1908) and 2007 in size classes dbh>45.7 cm, 45.7≥dbh≥10.2 cm, and 10.2 cm >dbh>2.5 cm for (**a**) *Castanea dentata*, (**b**) *Quercus rubra*, (**c**) *Quercus velutina*, and (**d**) *Prunus virginiana*

spinosa), sweet birch, and staghorn-sumac (*Rhus typhina*) were in all three forests; shagbark-hickory (*Carya ovata*) and sweet gum (*Liquidambar styraciflua*) were only in Beechwood and Country Club; and umbrella-tree (*Magnolia tripetala*) was just in Beechwood.

Among the species with an sph greater than two in 1868 (Fig. 2.4) and not reported above, sugar maple was not present in 1907 but in 2007 was found only in

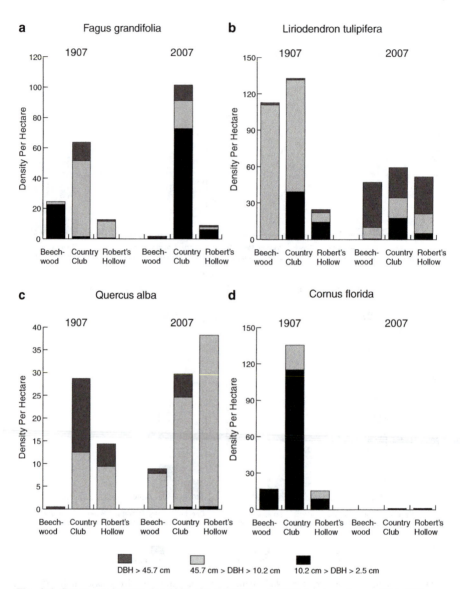

Fig. 2.6 Stems per hectare for 1907 (Paul 1908) and 2007 in size classes dbh>45.7 cm, 45.7≥dbh≥10.2 cm, and 10.2 cm>dbh>2.5 cm for (**a**) *Fagus grandifolia*, (**b**) *Liriodendron tulipifera*, (**c**) *Quercus alba*, and (**d**) *Cornus florida*

Beechwood and Country Club with an sph less than three. Tree of heaven (*Ailanthus altissima*) was reported in Country Club and Robert's Hollow in 1907 but remained only in Country Club in 2007 with an sph under three. Southern catalpa had an sph less than one and was present only in Robert's Hollow during 2007. Black walnut

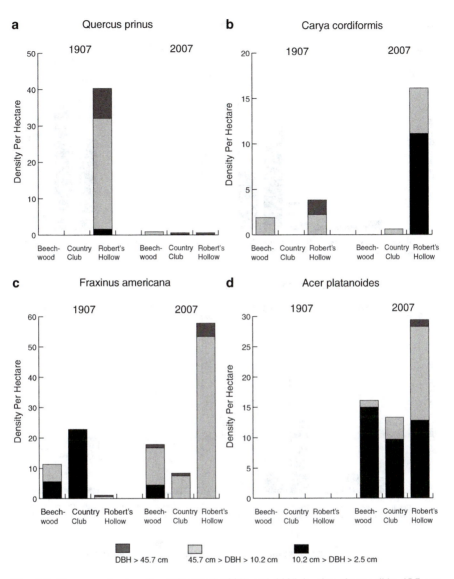

Fig. 2.7 Stems per hectare for 1907 (Paul 1908) and 2007 in size classes dbh>45.7 cm, 45.7≥dbh≥10.2 cm, and 10.2 cm>dbh>2.5 cm for (**a**) *Quercus prinus*, (**b**) *Carya cordiformis*, (**c**) *Fraxinus americana*, and (**d**) *Acer platanoides*

(*Juglans nigra*) was found in both 1907 and 2007 in Country Club and Robert's Hollow; however, the total sph did not exceed four. Sycamore had an sph less than two and was located just in Robert's Hollow during 1907. Three specimens of white pine were reported only in the sampling notes for Country Club in 1907 and were not present in the area of the forest during the 2007 field work. Black locust was first

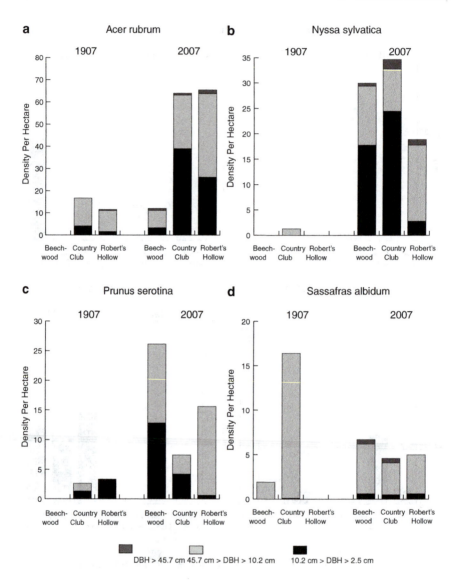

Fig. 2.8 Stems per hectare for 1907 (Paul 1908) and 2007 in size classes dbh>45.7 cm, 45.7≥dbh≥10.2 cm, and 10.2 cm>dbh>2.5 cm for (**a**) *Acer rubrum*, (**b**) *Nyssa sylvatica*, (**c**) *Prunus serotina*, and (**d**) *Sassafras albidum*

recorded in Beechwood and Robert's Hollow in 2007 with an sph less than one. Silver-maple, eastern red cedar, English walnut (*Juglans regia*), black mulberry, European bird-cherry (*Prunus padus*), and black willow were not found in the three forests in 1907 or 2007.

Forest Dynamics

Plantings of native trees in the estates that became Fairmount Park formed a forest dominated by native species in 1868 (Fig. 2.4) which reflected the pre-European settlement forest composition as indicated by witness tree records (Loeb 1987). In contrast, balsam fir, black spruce, eastern hemlock, red buckeye, and scarlet oak were commonly planted before 1800 (Adams 2004) but were not represented in Fairmount Park by specimens with dbh>45.7 cm (Cresson 1868). Paul (1908) pointed out that air pollution caused the loss of some gymnosperm species, which may explain the absence of balsam fir, black spruce, and eastern hemlock in 1868 (the field surveyor tally sheet used in 1868 included eastern hemlock but none were reported). All of the species with an sph greater than two in 1868 were found in the three forests except silver-maple, English walnut, eastern red cedar, black mulberry, European bird-cherry, and black willow. The large population of black mulberry in 1868 could have been a remnant plantation for the silk industry in Philadelphia (plantings occurred from 1769 to 1843; Scharf and Westcott 1884). The three forests are far from the lake or stream environments usually associated with silver-maple and black willow. English walnut was probably a minor component of the pairing with black walnut. European bird-cherry is a species that reproduces well in cutover Pennsylvania forests but is not competitive when the canopy redevelops (Elliott 1927). The loss of a large population of eastern red cedar also occurred in Seton Falls Park, New York City, NY (Loeb 1989b), and Alley Pond Park, New York City (Loeb 1992b).

The forest dynamics of the Beechwood, Country Club, and Robert's Hollow can be differentiated based upon their history of disturbances: Beechwood was periodically burned and experienced trampling from pedestrian traffic; Country Club was not affected by railroad-engine-related fires but park visitors and horseback riders had access to the site, and the forest had the dead trees removed in 1905; and Robert's Hollow did have severe fires but no regular human visitors. Chestnut, tulip-tree, choke-cherry, northern red oak, and black oak being the dominant species for Beechwood in 1907 could be explained by the fires but the question of how frequently did burning occur comes to the forefront because the species had little or no reproduction, and as Paul (1908) observed the sprouts were short lived. The absence of railroad fires is reflected in Country Club by the large population of flowering dogwood and the presence of saplings or larger sapling sph than found in Beechwood for chestnut, tulip-tree, choke-cherry, and black oak. In all three forests, chestnut, black oak, white oak, rock chestnut-oak, and northern red oak had few or no saplings, which could have been caused by shading from the understory composed of the flowering dogwood, choke-cherry, white ash, and red maple (Lorimer et al. 1994). Although chestnut, black oak, white oak, and northern red oak were well represented in Robert's Hollow, the large population of rock chestnut-oak was unique. Perhaps relatively little trampling in the secluded location of Robert's Hollow permitted rock chestnut-oak to reproduce and grow more quickly as compared to chestnut, black oak, white oak, and northern red oak.

During the century from 1907 to 2007, two changes affected all three forests: chestnut was lost, which opened the canopy soon after 1907, and white-tailed deer began to browse after 1980. The cessation of railroad fires by 1955 changed conditions only in Beechwood and Robert's Hollow (no evidence of fire was found in the three forests during the 2007 field work). The amount of trampling in Beechwood and Country Club was reduced by the end of trolley service in 1946. Aughanbaugh (1935) predicted from observations in old growth rural forests of southern Pennsylvania that the canopy openings left by the loss of chestnut would be filled with the fast growing species northern red oak and tulip-tree. The total sph for northern red oak in each of the three old growth forests in Fairmount Park is nearly equal, which is not the situation for the landscaped old growth urban forests of New York City (Stalter and Kincaid 2008). Tulip-tree nearly follows the pattern of northern red oak but the sapling sph for Country Club and Robert's Hollow is less in 2007 than 1907. Again, New York City forests have very different sph values for tulip-tree with lowland forests having the highest sph (Fitzgerald and Loeb 2008).

In 1907, white oak appeared to be poised to disappear from the forests but instead the 2007 data revealed expansions in the subcanopy size class in all three forests and the first reported presence of saplings in Country Club and Robert's Hollow. The increases in white oak differ from the declines in rural forests (Abrams 2003), which may be related to the higher levels of disturbance in the three Fairmount Park forests then occured in rural forests. Comparing 1985 to 2001 in the landscaped old growth urban forest of Inwood Hill Park, New York City (Fitzgerald and Loeb 2008), white oak trees and saplings had lower sph values; however, the total sph in 2001 was more than double the values for the forests of Fairmount Park in 2007. American beech almost disappeared from Beechwood by 2007 but the sph for saplings rose in both Country Club and Robert's Hollow. The American beech sapling increase is primarily from roots sprouts in reaction to root trampling (Busby et al. 2008) but also could be related to a preference against browsing by white-tailed deer (Krueger et al. 2009). A rise in American beech root sprouts occurred in Seton Falls Park (Loeb 1982) and the Hemlock Forest, New York Botanical Gardens, New York City (Rudnicky and McDonnell 1989) without the presence of white-tailed deer. The combination of pedestrian and horseback riding causes greater damage to tree roots (Landsberg et al. 2001) which may explain why American beech sprouting is more prevalent in Country Club than Robert's Hollow.

The shade-intolerant choke-cherry did not survive in the three forests and has not been reported in old growth urban forest studies in New York City (Stalter and Kincaid 2008). Black oak appears to be headed for loss from the Fairmount Park forests but continues to successfully reproduce in Inwood Hill Park (Fitzgerald and Loeb 2008). Flowering dogwood representation by just a few sapling size specimens in Country Club and Robert's Hollow in 2007 could have been caused by the recent dogwood anthracnose disease (*Discula destructiva* Redlin; Hibben and Daughtrey 1988). However, McCormick (1971) reported flowering dogwood was not present in Beechwood during 1969. Also, 90% of the flowering dogwood in Seton Falls Park was lost by 1979 (Loeb 1982), which is 4 years before the disease was found in New York and Pennsylvania. Red maple, Norway maple, Hercules club, sweet birch, hornbeam, bitternut-hickory, shagbark-hickory, white ash, sweet gum, umbrella-tree, black gum, wild black cherry, staghorn-sumac, and sassafras took advantage of the canopy openings created by the loss of chestnut to attain gains in sph or to be first reported in 2007. Umbrella-tree seedlings are thriving in Beechwood when seedlings for all of the other species including American beech are virtually absent because of deer browsing.

In New York City landscaped old growth urban forests, the 14 species noted above had sph increases except for umbrella-tree which is a species not reported in forest composition change studies (Fitzgerald and Loeb 2008; Stalter and Kincaid 2008). Many saplings of wild black cherry and sassafras occurred in repeatedly burned areas of Seton Falls Park (Loeb 1982) but the two species became established in the unburned Country Club forest. The most puzzling change in 2007 was the virtual loss of the former dominant species in Robert's Hollow, rock chestnut-oak, even though the species is well represented in the canopy, and sapling size classes of Inwood Hill Park (Fitzgerald and Loeb 2008). Paul (1908) reported "the white and chestnut oaks are suffering severely from a green scale (*Asterolecanium variolosum*), one of the most difficult and injurious insects we have encountered …. When we have been able to give the trees repeated sprayings, no serious damage has resulted but many trees not so protected have succumbed." In the northeastern United States, rock chestnut-oak is the most favored host of green scale with saplings being severely affected and mature trees dying when other factors stress the tree (Parr 1940). The rock chestnut-oaks may not have been treated because of the relative isolation of Robert's Hollow and the combination of fire and green scale could have caused the dramatic loss for rock chestnut-oak trees and saplings.

Remnant Forest

Agricultural Cessation

Original Research: Saddler's Woods, Haddon Township, New Jersey

Saddler's Woods (39° 54′ 8″ N, 75° 3′ 19″ W) was named after Joshua Saddler, an escaped slave who joined a Quaker farming community. In 1868, Saddler legally prevented cutting in the forest that developed on his family farm in New Jersey (Saddler's Woods Conservation Association 2010). The old growth urban forest was inventoried in 2011 with a 60 m by 60 m plot placed to have a minimum of a 10 m separation from the bordering disturbed forests. Saddler's Creek runs through the plot. Canopy gaps are present including the one caused by the fall of a large-diameter (>100 cm dbh) black oak during winter 2011. Among the 15 species identified in the sample (Table 2.4), only American beech, northern red oak, white ash, and white oak had trees in the canopy size class (dbh > 45.7 cm). American beech comprised 41% of the saplings (2.5 < dbh < 10.2 cm) and 37% of the subcanopy

Table 2.4 Stems per hectare for canopy trees, subcanopy trees, and saplings for Saddler's Woods, Haddon Township, New Jersey, in 2011

Species	Saplings 2.5<dbh <10.2 cm	Subcanopy 10.2≤dbh ≤45.7 cm	Canopy dbh >45.7 cm
Acer negundo	83	56	0
Acer rubrum	278	417	0
Betula papyrifera	0	28	0
Carpinus carolinana	56	83	0
Cornus florida	28	0	0
Celtis occidentalis	83	0	0
Fagus grandifolia	834	723	83
Fraxinus americana	56	222	0
Liquidambar styraciflua	0	28	0
Liriodendron tulipifera	139	0	111
Nyssa sylvatica	28	56	0
Quercus alba	139	56	167
Quercus prinus	28	0	0
Quercus rubra	111	56	111
Sassafras albidum	167	222	0

(10.2 cm≤dbh≤45.7 cm dbh) trees. Red maple, white ash, and sassafras had the second through fourth highest sph for subcanopy trees and no representation in the canopy.

Among the New York City historical forest ecology studies, the best comparison to Saddler's Woods is Seton Falls Park (Loeb 1992b) because Rattlesnake Creek runs through the Park and the site was a farmland before the American Revolution and then a protected woodland associated with a farm. The most striking difference is that the Saddler's Woods canopy sph in 2011 is more than 13 times greater than the Seton Falls Park canopy (dbh>50 cm) sph in 1979. Also, the Saddler's Woods subcanopy sph is almost 10 times greater than the subcanopy of Seton Falls Park (10.2 cm<dbh≤50 cm) in 1979. A visit by the author to Seton Falls Park in 2010 showed continued losses of trees from the canopy and subcanopy layers and little replacement occurring. Seton Falls Park had six taxa in 1979 not present at Saddler's Woods, American elm, black cherry, hemlock, black birch, black locust, and hickory (*Carya* spp.), while only elder maple (*Acer negundo*) was found in Saddler's Woods but not Seton Falls. The deep soils in Saddler's Woods could contribute to an explanation of the higher sph than found in the shallow soils of Seton Falls Park. Although vandals cut down trees in both Saddler's Woods and Seton Falls Park, evidence of fire was observed in the Park in 1979 and 2010 and is neither evident nor reported for Saddler's Woods. American beech having the largest sapling and subcanopy populations in Saddler's Woods appears to be related to preference against browsing by white-tailed deer (Krueger et al. 2009). Since many of the saplings and subcanopy trees appear to be root sprouts, the high population of American beech could be related, at least in part, to human trampling (Busby et al. 2008) as occurs in Seton Falls Park (Loeb 1982). Among the three Fairmount Park forests, Country Club forest is most similar to Saddler's Woods in terms of the distribution of stems among the canopy, subcanopy, and sapling sizes class. However, Saddler's Woods has twice

the total sph of the Country Club forest. Saddler's Woods and Country Club both have high American beech sapling populations apparently because of root sprouts and white-tailed deer preferences against browsing American beech. Red maple having the second highest number of sph for both saplings and the subcanopy in Saddler's Woods parallels the red maple population expansion in the Beechwood, Country Club, and Robert's Hollow forests.

Forest Cutting Cessation

The Mishow Marsh Watershed, Hunter Island, Pelham Bay, Park, Bronx, New York (40° 52′ 40″ N, 73° 47′ 4″ W) is the only remnant old growth urban forest with a known ecological history of a clear-cutting event and forest redevelopment (Loeb 1998). Paleopalynological research revealed the watershed was a native American corn field from 1105 to 1165 and then planted to become a hickory forest. The British Navy clear-cut Hunter Island in 1779, and subsequent plantings for estate development and park use of Hunter Island did not affect the watershed. From 1934 to 1989 tree (dbh > 30 cm) sph rose for hickory (2.6–11.2), oak (17.5–31.2), and sassafras (0–7.3), while apple (Pyrus spp.), American beech, black locust, and flowering dogwood were lost from the forest.

Limited Tree Harvesting

There are two remnant forests in the New York and Philadelphia metropolitan areas which are thought to be limited tree harvesting forests: the Hemlock Forest, New York Botanical Gardens, Bronx, New York (Rudnicky and McDonnell 1989), and the William L. Hutcheson Memorial Forest, New Jersey (Buell et al. 1954). The Hemlock Forest (40° 51′ 51″ N, 73° 52′ 34″ W) is located along the ravine of the Bronx River which provides the environmental conditions for hemlock to grow and reproduce. The forest has paved roads with water hydrants to enable responses to fires. Dead chestnut trees were cut and salvaged from the woods after the chestnut blight in 1908 (Pennsylvania Chestnut Tree Blight Commission 1912). From the 1880s to the 1910s, leaves were raked out of the forest which may have affected seedling survival. Rudnicky and McDonnell (1989) resampled the canopy (dbh ≥ 15 cm) with 227 plots (15% of the Hemlock Forest area) in order to compare the forest in 1985 to the inventory of the forest noted on the New York City Department of Parks and Recreation maps created in 1937. There was a small decrease in stems per hectare for the two dominants hemlock (52 sph in 1937 and 47 sph in 1985) and oak species (from 34 sph in 1937 to 31 sph in 1985). In contrast, there were large increases for red maple (4 sph in 1937 and 27 sph in 1985), black cherry (from 2 sph in 1937 to 14 sph in 1985), and white ash (1 sph in 1937 as compared to 16 sph in 1985). The history of trampling and annual fires causing loss of organic material in the soil of the Hemlock Forest that began with the founding of the Botanical Gardens in 1889 inhibits hemlock reproduction (Rudnicky and McDonnell 1989). Based on my personal observation in 2007 the canopy population of hemlock continues to decline in the Hemlock Forest, which appears to be related to the effects of the introduced hemlock woolly adelgid (*Adelges tsugae* Annand; Orwig et al. 2002). In Seton Falls Park, hemlock declined from 1936 to 1979 (Loeb 1982) and was lost by 2007.

The William L. Hutcheson Memorial Forest (40° 29″ 55″ N, 74° 33′ 48″ W) is a remnant forest associated with a farm but the forest was not cleared for agriculture. Although no record exists, presumably chestnut trees were lost to the chestnut blight. There are records of the losses of trees to a severe storm in 1950 and the downed timber was recovered for sale. Some of the downed trees had colonial-period fire scars (Buell et al. 1954). Wind is the predominant cause of tree losses with tree falls occurring more frequently at poorly drained sites than well drained sites. Oppositely, trees lost to wind snapping the trunk occurred more frequently in well-drained sites than poorly drained sites (Reiner and Reiner 1965). Overall, wind caused more trees losses in the poorly drained sites than the well-drained sites and the resultant canopy gaps permitted the advance of red maple and white ash into the canopy (Monk 1961a). In 1961, the number of seedlings per tree for sugar maple and Norway maple were far greater than the other species in the forest (Monk 1961b). Tree deaths related to a severe summer drought in 1957 were greatest for red maple, dogwood, and white oak (Small 1957). A study of tree seedling survivorship over a 15 month period (1979–1980) including a severe summer drought in 1980 showed that more than 70% of the Norway maple, hickory, white ash, black cherry, and white oak seedlings survived while less than 30% of the red maple, dogwood, and sweet cherry seedlings survived (Davison 1981).

Sulser (1971) examined change in forest composition by resampling (sampling plot locations approximated from a field notebook) 20 years after unpublished measurements in 1950 were taken prior to the storm that extensively damaged the forest. The only significant differences found were sapling increases for red maple and white ash. In 2003, Aronson (2007) resampled the plots measured in 1950 and 1970. Flowering dogwood tree importance value (IV) dropped from 149.3 in 1950 to 17.4 in 2003 and the sapling IV also dropped from 131.9 in 1950 to 0 in 2003. White oak tree IV declined from 69.6 in 1950 to 4 in 2003 and saplings were not present in either year. Black cherry tree IV rose from 0 in 1950 to 52.2 in 2003 while the saplings IV only increased from 3 to 9.3 over the same period. Red maple tree IV jumped from 4.5 in 1950 to 97.4 in 2003 but in contrast sapling IV was lower with 16.8 in 1950 and 9.3 in 2003. American beech trees and saplings disappeared from the sampling plots. Finally, Aronson (2007) found the combination of deer browsing and the invasive species stilt grass (*Microstegium vimineum*) caused regeneration failure for the canopy species and permitted the seedling populations of tree of heaven and white ash to expand.

Regional Synthesis of Historical Continuity

The paleopalynology record of the past century indicated five major changes affecting forest dynamics in the oak chestnut region: release of lands from agriculture, loss of chestnut, decimation of hickory, urban expansion instead of rural forest development, and urban tree plantings (Loeb 1989a). A paleopalynological study of one remnant old growth urban forest in New York City revealed pre-European settlement Native American plantings of hickory occurred soon after corn field abandonment (Loeb 1998). Floras (Barton 1818; Torrey 1819) published after the American Revolution, the regional forest resetting event, revealed 98 arboreal species with only 54% of the species being present in both the New York City and Philadelphia metropolitan areas. Even though alien species comprised only 5% of the total, more than 90% of the arboreal species were used for arboricultural plantings before the floras were published, which highlights the extensive use of native species in estate and garden plantings starting in the early colonial period (Fry 1996; Fitzgerald and Loeb 2008). Witness tree records (Loeb 1987) of pre-European settlement forests showed oak species were two-thirds of the forest and chestnut was the second dominant genus, which confirms the classification of the region as oak-chestnut forest. The most common oak species in the witness tree records were in descending order white oak, black oak, and northern red oak. The predominance of oaks indicates fire affected pre-European settlement forest, which is supported by the fire scar analysis of trees in the William L. Hutcheson Memorial Forest (Buell et al. 1954). However, twelfth-century native American plantings of hickory in the Mishow Marsh Watershed (Loeb 1998) points out the possibility of prehistoric arboricultural influences on species composition.

A treatise on street tree management and plantings in the New York and Philadelphia region (Solotaroff 1912) included the results of a 1911 street survey for East Orange, New Jersey. Three species, red maple, silver-maple, and sugar maple, comprised more than two-thirds of the 10,953 trees. By 2004, the street forest was down to 6605 trees with red, silver, and sugar maples being less than 10% of the total. Solotaroff indicated several problems with the growth of each species which made the trees unsuitable for urban conditions. In contrast, Solotaroff recommended Norway maple which had a slightly smaller population in 2004 than 1911 but became the second most common species behind pin oak. A comparison of surveys for the street old growth urban forests of Haddonfield and Moorestown in Camden County, New Jersey with the East Orange results indicates the species diversity of the two forests was more than three times greater than East Orange. Criteria for the assessment of whether a particular forest is a street old growth urban forest are the largest size class (dbh > 45.7 cm) has the highest sph for all species and the species with the greatest sph match the species in historical street forests (e.g., East Orange in 1911) or historical floras (such as found in Table 2.1). Haddonfield matched the criteria well but Moorestown did not which revealed how tree diseases and planting plans to increase diversity affect patterns in the historical continuity of street old growth urban forests.

In the three landscaped forests of Fairmount Park, the losses for chestnut, flowering dogwood, and choke cherry permitted American beech, black gum, northern red oak, Norway maple, red maple, sassafras, tulip-tree, and white oak to establish sapling populations and have increased presence in the subcanopy and canopy layers. The Robert's Hollow forest's former dominant species rock chestnut-oak appears to have been lost to intense fire and green scale infestation, which permitted black cherry, black locust, Norway maple, and red maple to become more important in the forest. Studies of old growth landscaped forests in New York City parks (Loeb 1982; Fitzgerald and Loeb 2008; Stalter and Kincaid 2008) indicate rock chestnut-oak survives and reproduces in the presence of ground fires. However, the fires permitted black cherry, black locust, Norway maple, and red maple to become more important in the sapling, subcanopy, and canopy layers of the New York City forests. In contrast, the absence of fire in a Cleveland, Ohio, area old growth urban forest has resulted in dominance by sugar maple (not found in the three Fairmount Park forests) and American beech (Loeb 2001). American beech reproduction was greatest in the Country Club forest in 2007 because of root sprouting in response to trampling by people and horses. Although Beechwood forest has virtually no American beech trees, deer browsing appears to have poised umbrella-tree saplings to advance in the subcanopy and canopy without competition. In the New York City old growth landscaped forests noted above, American beech reproduces well through root sprouting in the absence of white-tailed deer and umbrella-tree.

The Mishow Marsh Watershed Forest recovered after a clear-cut in 1779 but more important to the forest recovery process was the planting of hickory by native Americans in the twelfth century. Hickory and oak increased in the forest canopy from 1934 to 1989 but sassafras appeared and became a canopy species since 1934 because of arson (Loeb 1998). American beech was lost from the forest, which could be related to the absence of white-tailed deer on the island. For Saddler's Woods, New Jersey, white-tailed deer and human trampling have led to American beech becoming the dominant species among saplings and subcanopy trees. Also white ash, red maple, and sassafras have advanced in the sapling and subcanopy layers but evidence of fire is not obvious in Saddler's Woods.

Researchers in the two limited cutting remnant old growth urban forest, the Hemlock Forest (Rudnicky and McDonnell 1989) and William L. Hutcheson Memorial Forest (Aronson 2007), predicted the oak species dominance of the canopy will be lost to black cherry, red maple, and white ash, as well as the possibility of tree of heaven becoming part of the canopy. However, the reasons for the changes in each forest are different. In the William L. Hutcheson Memorial Forest, regeneration is succumbing to deer browsing and stilt grass invasion, but neither of these species is a concern for the Hemlock forest. Instead, the loss of canopy species regeneration is due to trampling and periodic fire destroying the organic layer of the soil. An explanation for the changes in the Hemlock forest comes from studies of the seed bed germination in urban soils. In two New York City parks, high germination rates were found for tree of heaven, black birch, sweet gum, and tulip tree but there was no evidence of seed germination for maple species, oak species, white ash, flowering dogwood, and sassafras was found (Kostel-Hughes et al. 1998b). Also, Kostel-Hughes et al. (1998a) found that small-seed trees (red maple, sweet gum, tree of heaven, tulip-tree, white ash, witch hazel, sugar maple, sassafras, black cherry, and striped maple) germinated in urban soil leaf litter that was less than half the depth of the urban leaf litter in which large-seed trees (species of oak and hickory) germinated.

The forest resetting event for the old growth urban forests of the New York and Philadelphia metropolitan areas was the American Revolution. Witness tree records provided evidence for the existence of the oak-chestnut forest region prior to the war and the historical floras indicated intraregional variation in species distribution and frequency of occurrence. Arboricultural use of native species as early as 1600 and through 1850 brings into question the origin of trees in old growth urban forests. Are the trees from arboricultural plantings or spontaneous regeneration of local trees? Street old growth urban forests shift from dominance by species of maple at the turn of the twentieth century to dominance by oak, linden, and plane trees at the start of the twenty-first century. Landscaped forests composed of native and alien species, such as the Centennial Fairgrounds in Fairmount Park, have had large species diversity losses. The historical continuity of landscaped and remnant forests dominated by native species was disrupted by species losses or tree population drops to the chestnut blight, Dutch elm disease, hickory bark beetle, hemlock woolly adelgid, and green scale. Choke cherry was lost to shading by canopy trees. Major population declines for flowering dogwood are not apparently related to known diseases or insect infestations. Fire from railroad engine sparks and vandals has enabled the advance of black cherry, red maple, black locust, and sassafras. Human trampling and white-tailed deer browsing has decimated the seedlings and sapling populations except for American beech and umbrella-tree which are the species that white-tailed deer avoid browsing. Invasive species are not yet a widespread threat in old growth urban forest but there is great potential for the displacement of native tree species (Loeb 2009). The effects of tree deaths and sapling losses are not uniform across the landscaped and remnant forest sites in the New York and Philadelphia region as evidenced by an order of magnitude difference for canopy and subcanopy stems per hectare between Saddler's Woods, New Jersey, and Seton Falls Park, New York.

Looking into the future, research on more sites in the Philadelphia and New York metropolitan areas is required for statistical analysis of forest dynamics on the regional level. The focus of historical continuity research needs to be expanded to include abiotic factors such air pollution, climate warming, and progressive soil fertility exhaustion through the analysis of tree ring wood. Monitoring of sites is essential to maintain the flow of data on the changing biotic and abiotic conditions in the old growth urban forests. Finally, research on old growth urban forest must be done across the earth in order to gain an understanding of whether the factors affecting historical continuity in the oak-chestnut forest region have parallels in other forest regions of the planet.

References

Abrams MD (2003) Where has all the white oak gone? Bioscience 53:927–939
Adams DW (2004) Restoring American gardens. Timber Press, Portland, Or
Agnoletti M, Anderson S (2000) Methods and approaches in forest history. CABI Publishing, New York, NY

Anonymous (1868) Map of farms and lots embraced within the limits of Fairmount Park as appro-
    priated for public use by act of Assembly approved 14th day of April A.D. 1868. No publ,
    Philadelphia, PA
Anonymous (1916) The dying hickories on Long Island. Branch For Shade Tree Insects, Bur
    Entomology, US Dep Agri, Washington DC
Anonymous (2006) Comprehensive master plan for East Orange, New Jersey. Unpubl, East
    Orange, NJ
Aronson MF (2007) Ecological change by alien plants in an urban landscape. Diss, Rutgers Univ,
    New Brunswick, NJ
Aughanbaugh JE (1935) Replacement of the chestnut in Pennsylvania. Bull 54, Pa Dep For Waters,
    Harrisburg, Pa
Baker CA, Moxey PA, Oxford PM (1978) Woodland continuity and change in Epping Forest. Field
    Stud 4:646–669
Bard GE (1952) Secondary succession on the Piedmont of New Jersey. Ecol Monogr 22:196–215
Barton WPC (1818) Compendium floræ Philadelphicæ. Vols 1–2. M. Carey and Son,
    Philadelphia, Pa
Behre K-E (ed) (1986) Anthropogenic indicators in pollen diagrams. Balkema, Rotterdam, Neth
Birks HH, Birks HJB, Kaland PE, Moe D (eds) (1988) The cultural landscape: past, present, and
    future Cambridge Univ Press, Cambridge, UK
Brande A, Böcker R, Graf A (1990) Changes of flora, vegetation and urban biotopes in Berlin
    (west). In: Sukopp H, Hejný S (eds) Urban ecology. Plants and plant communities in urban
    environments. SPB Academic Publishing, The Hague, Neth
Braun EL (1950) Deciduous forests of eastern North America. Hafner, New York, NY
Brown CS (2005) Finding of no significant impact and decision. Environmental assessment –
    shooting white-tailed deer to assist the city of Philadelphia, Fairmount Park Commission in
    achieving deer population reductions on park properties located in the Pennsylvania counties
    of Delaware, Montgomery and Philadelphia. US Dep Agric, Anim Plant Health Insp Serv,
    Wildl Serv, Harrisburg, Pa
Buell MF, Buell HF, Small JA (1954) Fire in the history of Mettler's Woods. Bull Torrey Bot Club
    81:253–255
Busby PE, Motzkin G, Foster DR (2008) Multiple and interacting disturbances lead to *Fagus
    grandifolia* dominance in coastal New England. J Torrey Bot Soc 135:346–359
Cho D, Boerner REJ (1991) Structure, dynamics, and composition of Sears Woods and Carmean
    Woods State Nature Preserves, north-central Ohio. Castanea 56:77–89
Collins S (1956) The biotic communities of the Greenbrook Sanctuary. Diss, Rutgers Univ, New
    Brunswick, NJ
Corson A (1937) Report of landscape gardener for the year of 1936. In: (no ed) Fairmount Park
    annual report of the chief engineer for the year 1936. Fairmount Park Com, Philadelphia, Pa
Cox HE (1970) The Fairmount Park trolley, a unique Philadelphia experiment. Harold Cox, Forty
    Fort, Pa
Cresson J (1868) Report of the chief engineer of Fairmount Park. In: (no ed) Second annual report
    of the Fairmount Park Commission. Fairmount Park Com, Philadelphia, Pa
Crowder MJ, Hand DJ (1990) Analysis of repeated measures. Chapman and Hall, London, UK
Davison SE (1981) Tree seedling survivorship at Hutcheson Memorial Forest. William L Hutcheson
    Meml For Bull 6:4–7
DeCandido RV, Lamont EE (2004) The historical and extant vascular flora of Pelham Bay Park,
    Bronx county, New York 1947–1998. J Torrey Bot Soc 131:368–386
Dimbleby GW (1985) The palynology of archaeological sites. Academic Press, Orlando, Fl
Egan D, Howell EA (eds) (2001) The historical ecology handbook: a restorationists' guide to refer-
    ence ecosystems. Island Press, Washington, DC
Elliott HE (1927) What follows pulp and chemical wood cutting in northern Pennsylvania. Bull 43
    Spec Stud Ser, Comm Pa, Dep For Waters, Harrisburg, Pa
Fairmount Park Commission (1868) Fairmount Park, Philadelphia with limits, as prescribed in Act
    of Assembly, approved March 26th, 1868 showing the trees and woods nearly as now existing
    with a study for roads and paths. Worley & Bracher, Philadelphia, Pa

Fairmount Park Commission (1895) Proposal for cutting grass in Fairmount Park for the season of 1895. MC 1999.13, Eli Kirk Price Family Papers, Arch Spec Collect, Dickinson College, Carlisle, Pa

Fairmount Park Commission (1900) Topographical map of Fairmount Park Philadelphia excepting Wissahickon Valley. Fairmount Park Com, Philadelphia, Pa

Fitzgerald JM, Loeb RE (2008) Historical ecology of Inwood Hill Park, Manhattan, New York. J Torrey Bot Soc doi:10.3159/07-RA-046.1

Fry JT (1996) An international catalogue of North American trees and shrubs: the Bartram broadside, 1783. J Gard Hist 16:3–66

Gibson J, Eck RO, Jones T, Sterling R, Hartman J, Hannum W, Nichols RE, Thomas W, Leusner B (2010) Community forestry management plan 2010–2014 township of Moorestown. Unpubl, Moorestown, NJ

Gleason HA (1939) The individualistic concept of the plant association. Am Midl Nat 21:92–110

Gleason HA, Cronquist A (1991) Manual of vascular plants of the northeastern United States and adjacent Canada. New York Bot Gardens, New York, NY

Graney DL (1987) Ten-year growth of red and white oak crop trees following thinning and fertilization in the Boston Mountains of Arkansas. In: Proceedings of the fourth biennial southern silvicultural research conference. Gen Tech Rep SE-42. SE For Exp Stat, For Serv, US Dep Agri, Asheville, NC

Greller AM (1972) Observations on the forests of northern Queens county, Long Island, from colonial times to the present. Bull Torrey Bot Club 99:202–206

Hannah PR (1968) Topography and soil relations for white and black oak in southern Indiana. Res Pap NC-25. NC Res Stat, For Serv, US Dep Agri, St. Paul, Mn

Hepting GH (1977) The threatened elms: a perspective on tree disease control. J For His 21:90–97

Hibben CR, Daughtrey ML (1988) Dogwood anthracnose in northeastern United States. Plant Dis 72:199–203

Hoglund AW (1962) Forest conservation and stove inventors-1789–1850. J For His 5(4):2–8

Holton JL (1989) The Reading Railroad: history of a coal age empire. Garrigues House, Laury's Station, Pa

Kays R, Wilson D (2009) Mammals of North America. Princeton Univ Press, Princeton, NJ

Keller IA, Brown S (1905) Handbook of the flora of Philadelphia and vicinity, containing data relating to the plants within the following radius: eastern Pennsylvania; all of New Jersey except the northern counties; and New Castle County, Delaware, with keys for identification of species. Philadelphia Bot Club, Philadelphia, Pa

Kostel-Hughes F, Young TP, Carreiro MM (1998a) Forest leaf litter quantity and seedling occurrence along an urban-rural gradient. Urb Ecosyst doi:10.1023/A:1009536706827

Kostel-Hughes F, Young TP, McDonnell MJ (1998b) The soil seed bank and its relationship to the above ground vegetation in deciduous forests in New York City. Urb Ecosyst doi:10.1023/A:1009541213518

Krueger LM, Peterson CJ, Royo A, Carson WP (2009) Evaluating relationships among tree growth rate, shade tolerance, and browse tolerance following disturbance in an eastern deciduous forest. Can J For Res doi:10.1139/X09-155

Landsberg J, Logan B, Shorthouse D (2001) Horse riding in urban conservation areas: reviewing scientific evidence to guide management. Ecol Manag Restor doi:10.1046/j.1442-8903.2001.00067.x

Loeb RE (1982) Reliability of the New York City Department of Parks and Recreation forest records. Bull Torrey Bot Club 109:537–541

Loeb RE (1987) Pre-European settlement forest composition of East New Jersey and southeast New York. Am Midl Nat 118:414–423

Loeb RE (1989a) Lake pollen records of the past century. Palynology 13:3–19

Loeb RE (1989b) Historical ecology of an urban park. J For Hist 33:134–143

Loeb RE (1990) Measurement of vegetation changes through time by resampling. Bull Torrey Bot Club 116:173–175

Loeb RE (1992a) Will a tree grow in Brooklyn? Developmental trends of the New York City street tree forest. J For 90(1):20–24

Loeb RE (1992b) Long-term human disturbance of an urban park forest, New York City. For Ecol Manag doi:10.1016/0378-1127(92)90142-V

Loeb RE (1998) Urban forest management and ecosystem change during the past millennium: a case study from New York City. Urb Ecosys doi:10.1023/A:1009545331265

Loeb RE (2001) Fire in the urban forest: long-term effects in old growth stands. Arboric J 25:307–320

Loeb RE (2009) Biogeography of invasive plant species in urban park forests. In: Kohli R, Jose S, Batish D, Singh H (eds) Invasive plants and forest ecosystems. CRC/Taylor and Francis, London, UK

Loeb RE (2010) Diversity gained, diversity lost: long-term changes in woody plants in Central Park, New York City and Fairmount Park, Philadelphia. Stud Hist Gard Des Landsc doi:10.1080/14601170903040819

Lorimer CG, Chapman JW, Lambert WD (1994) Tall understorey vegetation as a factor in the poor development of oak seedlings beneath mature stands. J Ecol 82:227–237

Marx TG (1976) Technological change and the theory of the firm: the American locomotive industry 1920–1955. Bus Hist Rev 5:1–24

McBride JR, Nowak DJ (1989) Urban park tree inventories. Arboric J 13:345–361

McCarthy BC, Quinn JA (1990) Reproductive ecology of *Carya* (*Juglandaceae*): phenology, pollination, and breeding systems of two sympatric tree species. Amer J Bot 77:261–273

McCormick J (1971) An ecological inventory of the West Park, Fairmount Park, Philadelphia, Pennsylvania. Jack McCormick and Associates, Philadelphia, Pa

Miller JA (1965) The changing forest: recent research in the historical geography of American forests. J For His 9:18–25

Monk CD (1961a) The vegetation of the William L. Hutchison Memorial Forest, New Jersey. Bull Torrey Bot Club 88:156–166

Monk CD (1961b) Past and present influences on reproduction in the William L. Hutcheson Memorial Forest, New Jersey. Bull Torrey Bot Club 88:167–175

Nriagu JO (ed) (1978) The biogeochemistry of lead in the environment Part A. Ecological Cycles. Elsevier, Amsterdam, Neth

Orwig DA, Foster DR, Mausel DL (2002) Landscape patterns of hemlock decline in New England due to the introduced hemlock woolly adelgid. J Biogeogr doi:10.1046/j.1365-2699.2002.00765.x

Parr T (1940) *Asterolecanium variolosum* Ratzeburg, a gall-forming Coccid, and its effect upon the host trees. Number 40, Bull Yale Sch For

Paul O (1908) Report on the trees of Fairmount Park; a study of the trees growing naturally in the park forests and of those planted for shade or decorative purposes, including the outline of a general forestry policy suggested for their future care. Fairmount Park Com, Philadelphia, Pa

Peglar SM, Fritz SC, Birks HJB (1989) Vegetation and land-use history at Diss, Norfolk, U.K. J Ecol 77:203–222

Pennsylvania Chestnut Tree Blight Commission (1912) The chestnut blight disease. Bull 1, C E Aughinbaugh, Harrisburg, Pa

Peter D (2008) Tree succession planning: modelling tree longevity in Tuttangga/Park 17, the Adelaide park lands. Diss, Univ Adelaide, Adelaide, Aust. http://digital.library.adelaide.edu.au/dspace/handle/2440/48538. Accessed 12 March 2011

Peterken GF, Backmeroff C (1988) Long-term monitoring in unmanaged woodland nature reserves. Res Surv Nat Conserv Ser 9, Nat Conserv Counc, Peterborough, UK

Rackham O (2003) Ancient woodland its history, vegetation and uses in England, 2nd edn. Castlepoint Press, Colvend, UK

Rehder A (1940) Manual of cultivated trees and shrubs hardy in North America: exclusive of the subtropical and warmer temperate regions, 2nd edn. Macmillan, New York, NY

Reiner NM, Reiner WA (1965) Natural harvesting of trees. William L. Hutcheson Meml For Bull 2:9–17

Rotherham ID, Jones M, Smith ML, Handley C (eds) (2008) The woodland heritage manual a guide to investigating wooded landscapes. Wildtrack Publishing, Sheffield, UK

Rothrock JT (1880) Catalogue of trees and shrubs native of and introduced in the horticultural gardens adjacent to Horticultural Hall, in Fairmount Park. No publ., Philadelphia, Pa

Rudnicky JL, McDonnell MJ (1989) Forty-eight years of canopy change in a hardwood-hemlock forest in New York City. J Torrey Bot Soc 116:52–64

Russell EWB (1979) Vegetational change in northern New Jersey since 1500 A.D.: a palynological, vegetational, and historical synthesis. PhD thesis, Rutgers Univ, New Brunswick, NJ

Saddler's Woods Conservation Association (2010) History. http://www.saddlerswoods.org/35807.html. Accessed 21 March 2011

Scharf JT, Westcott T (1884) History of Philadelphia, 1609–1884. Everts, Philadelphia, Pa

Seppä H (1997) The long-term development of urban vegetation in Helsinki, Finland: a pollen diagram from Töölönlahti. Veg Hist Archaeobotany doi:10.1007/BF01261957

Sidney JC, Adams A (1859) Description of plan for the improvement of Fairmount Park. Merrihew and Thompson, Philadelphia, Pa

Small JA (1957) Drought response in William L. Hutcheson Memorial Forest. Bull Torrey Bot Club 88:180–183

Solomon AM, Kroener DF (1971) Suburban replacement of rural land uses reflected in the pollen rain of northeastern New Jersey. Bull NJ Acad Sci 16:30–44

Solotaroff W (1912) Shade-trees in town and cities their selection, planting, and care as applied to the art of street decoration; their diseases and remedies; their municipal control and supervision. John Wiley & Sons, New York, NY

Stalter R, Kincaid D (2008) A 70-year history of arborescent vegetation of Inwood Park, Manhattan, New York, U.S. Arboric Urban For 34:245–251

Sulser JR (1971) Twenty years of change in the Hutcheson Memorial Forest. William L Hutcheson Meml For Bull 2:15–25

Taylor N (1915) Flora of the vicinity of New York: a contribution to plant geography. Vol 5 Memoirs New York Bot Gard, New York, NY

Torrey J (1819) Catalogue of plants growing spontaneously within thirty miles of the city of New-York. Websters and Skinners, Albany, NY

Traverse AT (2007) Paleopalynology, 2nd edn. Springer, London, UK

Vera FWM (2000) Grazing ecology and forest history. CABI Publishing, New York, NY

Ward JS, Parker GR, Ferrandino FJ (1996) Long-term spatial dynamics in an old-growth deciduous forest. For Ecol Manag 83:189–202

Chapter 3
Partnerships, Adaptive Management, and Restoration for Historical Continuity

Abstract The future of old growth urban forests is not solely in the hands of the individuals or government agencies that at some time in the past had sufficient funds to plant and maintain the forests. Partnerships of landowners; local government; nearby schools, colleges, and universities; and community neighbors must collaborate to muster the resources needed to assure the historical continuity of old growth urban forests. Because threats such as invasive species, deer browsing, and human trampling are poorly understood, partnerships utilize adaptive management in order to respond to the challenges facing old growth urban forests. Foremost among partnership activities is communication to help forest visitors understand the need for maintenance and restoration activities. Plans for historical continuity restoration projects are based on records of forest changes revealed by the historical ecology research on the individual forest. The spatial limitations of street and landscaped forests leads to the planting of microforests with canopy, subcanopy, and ground-level species placed within the confines of the available growth environment. In addition to plantings, restoration of remnant forests requires restricted access to the forest in order to eliminate disturbances which kill the spontaneous regeneration of arboreal species essential to the reestablishment of forest dynamics and historical continuity.

Keywords Adaptive management partnership • Historical continuity restoration • Microforests • Invasive species • Public communication • Human trampling • Deer browsing

Partnerships

Although laws exist to protect urban trees around the globe (Profous and Loeb 1990), economic conditions affect government funding for the management and restoration of urban forests (Loeb 1987). When funds are insufficient to meet the needs of urban

Fig. 3.1 Management by naturalization leading to forest loss over the long term (Courtesy of the Green Seattle Partnership – City of Seattle Contact Mark Mead)

forests, the alternative selected by government agencies is naturalization, a virtually no-cost process in which maintenance and restoration is stopped and access to the area is supposed to be limited (Millward et al. 2011). The problem with naturalization is the historical continuity of the old growth urban forest is ignored and will be lost to disturbances (Fig. 3.1). The alternative to management of old growth urban forest by government agencies alone is the formation of partnerships to maintain and restore the forest. Why should neighbors join together in efforts to maintain the historical continuity of an old growth urban forest in the community? The answer is based in the fact that more than half of the world's population lives in urban areas (Müller and Werner 2010). Old growth urban forest is the primary experience of a forest for most people. An old growth urban forest opens the door for city people to be inspired by forests, to interpret forest ecology, and to be motivated to take action in support of the management and restoration of forests. Considering the importance of urbanites to the future of forests in urban and rural areas, restoring and maintaining the historical continuity of old growth urban forests will have a broad positive effect for all forests.

The value of involving community neighbors in old growth urban forest management and restoration is enhanced decision-making, development of a sense of ownership of the restoration process, and raised awareness of issues related to the future of the forest (Van Hertzele et al. 2005). The undesirable alternative is uncoordinated actions by individuals which are certainly destined to cause harm to an old growth urban forest, at least over the long term. A resilient broad-based partnership with sensitive leaders is needed to systematically address the complex challenges affecting an old growth urban forest (Westphal and Childs 1994; Linnea 2010).

The partnership for an old growth urban forest should be composed of at least four contributory groups: the landowners; local government; local schools, colleges, and universities; and community neighbors. The landowner is vital to be on board in order to provide access to the forest and to assure that the forest will be protected and not lost to development. Local government may be able to provide protection from development through recognition of the site and law enforcement agency support in preventing criminal activities including vandalism. The teachers and professors of educational institutions provide scientific expertise and more importantly involve students in conducting the vital scientific research that guides management and restoration efforts. Students learn to value and protect the old growth urban forest and become community neighbors who are motivated to work for and lead the partnership. Leaders of the partnership have the responsibility for maintaining organizational continuity; helping individuals to take ownership of management, restoration, fund raising, and communication tasks; and advocating for the forest. Regional organizations can spawn partnerships for individual forests. For example, the Green Seattle Partnership has implemented the Tree Ambassador program which focuses on developing stewards of the urban forest who serve as resources for their local community. The Tree Ambassadors training program includes the basics of urban forestry, leadership, and community organizing (Green Seattle Partnership 2011a).

The two goals of an old growth urban forest partnership are adaptive management and restoration of historical continuity. The first step is translating the goals into action objectives. Each action objective must demonstrate the shared value of the forest in order to motivate partnership members to take on responsibilities. The most important action objective is communication and public outreach because involving community neighbors with the efforts for the old growth urban forest is essential to the success of the partnership. Clear communication with the public avoids misinterpretation of plans and differences in perceptions of specific restoration actions (Stein and Moxley 1992; Raffetto 1993; Gobster 1997). Also, special attention must be given to the ethnic make-up of the community because research into cultural differences in response to urban forests suggests that urban African-Americans have a preference for open woodland over dense woodlands with extensive undergrowth (Elmendorf et al. 2005). Effective communication engages and educates the local community with a common language that describes adaptive management and historical continuity restoration efforts. Obtaining feedback from the community and checking whether the chosen lexicon is successful in sharing meaning beyond the theoretical level are essential components of effective communication. If a communication strategy or campaign has attained a level of success then the public has gained an understanding of why individuals are to act in order to manage and restore old growth urban forests. The bottom line benefit for the partnership is motivation of the public to support activities sponsored by the partnership through personal actions and donations of cash, supplies, and services (Jones et al. 2005). The ideal outcome of a communication program is a "Smokey the Bear" success story in which the issues and needs of old growth urban forests are related to individual actions at the level of popular culture. One example of a public oriented communication strategy showing why urban forest restoration is vital was developed by the Cascade Land Conservancy and the Green Seattle Partnership. Two diagrams illustrate the current condition of urban forests in Seattle and what will happen to the forests if restoration does not occur (Fig. 3.2) versus the alternative of performing restoration (Fig. 3.3).

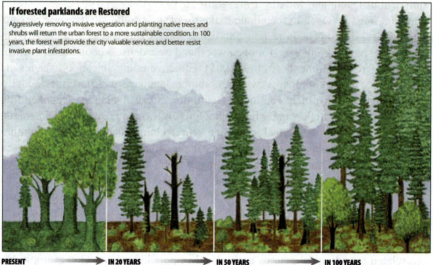

If forested parklands are Not Restored

Aggressive non-native vegetation will dominate the urban forest unless removed. In 100 years, the trees will be gone.

PRESENT

Forested parklands are dominated by deciduous trees, mainly big-leaf maples and alders, nearing the end of their life. After decades of neglect, non-native invasive plants, such as English ivy and wild clematis, cover the ground and grow up into the tree canopy.

IN 20 YEARS

Invasive plants outcompete and grow over existing native vegetation, blocking the sunlight plants and trees need to thrive. English ivy now dominates the tree canopy, making the trees weak, top heavy and susceptible to windfall. Eventually, trees die or fall over.

IN 50 YEARS

The trees are gone. Only a few native shrubs struggle to survive the stress of competition with invasive plants.

IN 100 YEARS

The forest is destroyed. Native trees can no longer establish on their own. We are left with a dense "ivy desert." Very few plant species can live, and forest biodiversity is gone. Such conditions provide homes for rats and scarce habitat for more desirable urban wildlife.

Fig. 3.2 Long-term effects of no forest restoration on the urban forest of Seattle, Washington (Courtesy of the Green Seattle Partnership – City of Seattle Contact Mark Mead)

If forested parklands are Restored

Aggressively removing invasive vegetation and planting native trees and shrubs will return the urban forest to a more sustainable condition. In 100 years, the forest will provide the city valuable services and better resist invasive plant infestations.

PRESENT

Forested parklands are dominated by deciduous trees, such as big-leaf maples and alders, nearing the end of their life. After decades of neglect, non-native invasive plants such as English ivy are smothering native vegetation and weakening native trees.

IN 20 YEARS

Through restoration efforts and long-term maintenance, the non-native plants are removed. Native groundcovers, shrubs and evergreen trees such as Douglas firs and Western red cedars and hemlocks are planted.

IN 50 YEARS

As the evergreen trees grow, they shade out sun-loving invasive plants such as blackberry. Native understory plants thrive.

IN 100 YEARS

With continued stewardship, the maturing forest requires less care and provides greater benefits to the city.

Fig. 3.3 Long-term effects of forest restoration on the urban forest of Seattle, Washington (Courtesy of the Green Seattle Partnership – City of Seattle Contact Mark Mead)

The most important communication strategy is providing educational opportunities for children and youth that are comprised of interactions with an old growth urban forest in a complementary fashion to traditional education. For example, the Green Seattle Partnership has developed the Urban Forestry Project (Andrea Mojzak, Personal Communication) which provides curricula for students at the elementary, middle, and high school levels. In addition to increasing the students' knowledge concerning the forest, the curriculum involves the students with taking actions to address threats to the forest such as invasive plants. Implementation of school curricula adapted to local forest conditions is essential because the future of old growth urban forest depends upon young people becoming involved in the actions needed to maintain and restore the historical continuity of the forest.

Adaptive Management

Adaptive management is the preferred model when more is unknown than known concerning forest management (McFadden et al. 2011), which is certainly the situation for old growth urban forests. Application of the adaptive management model calls for the identification of partnership concerns as the starting point for planning. Each concern is translated into an action objective (i.e., fund raising, invasive species removal, etc.) to be achieved through multiple events. In an iterative process, a method to address an action objective is implemented and progress toward achievement of the expected outcome is assessed. Repetition with correction of the method or the substitution of new methods in subsequent iterations contributes to outcome achievement.

The first step in developing an adaptive management plan is to conduct a forest threat assessment. Potential action objectives are derived from the assessment. Preparation of a prioritized list of action objectives by a small group obviates the need for long discussions of action objectives that may not be implemented. Next, the leaders must be prepared to explain the implementation processes for the action objectives in order to gain consensus on the adaptive management plan for the partnership. Once consensus is achieved, the management plan must be written as an outline of action objectives based on scientific evidence of need and described in a fashion that motivates involvement by the reader. Examples of common action objectives (with research citation to provide scientific evidence given in parenthesis) are as follows: maintain the forest by building trails to limit disturbance across the forest (Lehvävirta 1999), decrease erosion by constructing soil retention devices (Lehvävirta et al 2004), and restore forest dynamics by planting native species to produce a range of age classes for each tree species (Lehvävirta and Rita 2002).

Invasive species are an important early management priority for partnerships because invasive plants and animals (Maitland 1995) are constant and pervasive threats to old growth urban forests (Loeb 2009; Zipperer 2010). Adaptive management of invasive species is based on the implementation of strategies designed to achieve the action objective derived from the invasive species threat. Several strategies are needed because invasive species removal processes have to be expanded with modifications for years (Vidra et al. 2007; Loeb et al. 2010). Tu and Meyers-Rice (2001) prepared an invasive species remediation plan format that serves as a guide for planning iterative invasive plant removals and native species plantings to occupy the vacant niche space resulting from the removals. The processes for invasive species removal and native species planting have been described in a guide developed for Seattle (Green Seattle Partnership 2011b) but the document is broadly applicable to other metropolitan areas.

The experiences of the Saddler's Woods Conservation Association, New Jersey (Janet Goehner-Jacobs Personal Communication), with invasive species removal treatments and plantings of native species to fill vacant niches provide an insightful example. The disturbed forest surrounding the old growth urban forest in Saddler's Woods contains several invasive species (Fig. 3.4). The first iteration of the invasive species strategy in 2002 was manual removal of a wall of Japanese knotweed (*Polygonum cuspidatum*) and English ivy (*Hedera helix)* in a portion of the disturbed forest. In light of the success achieved despite conventional wisdom against manual removal, a second round of manual removal in 2002–2004 included digging up roots and planting native species. For the third iteration, professional herbicide applications were done in 2005 after bags were placed over native species. The young knotweed plants were reduced by nearly 90% without damage to the native plants. An invasive species transition occurred during 2005–2006; a garlic mustard (*Alliaria petiolata*) population became established which was treated with hand removal and native species plantings as well as spot spraying of knotweed. During 2006–2008, a second invasive species transition involved bedstraw (*Galium boreale*) and sticky willy (*G. aparine*) crowding out garlic mustard. Seeds carried into Saddler's Woods on the boots of volunteers performing the invasive species removal work are thought to be the origin of a stilt grass (*Microstegium vimineum*) population becoming established along the edges of the treatment site. During the next year (2008–2009), garlic mustard decreased with spontaneous growth of native violets (*Viola* spp.), Jack in the pulpit (*Arisaema triphyllum*), and tulip-tree. Also, well dispersed Japanese angelica (*Aralia elata*) and white mulberry (*Morus alba*) plants appeared and were removed by hand. During 2009–2010, there was a dramatic increase in Japanese angelica with some trees being so large as to require removal by hand because the leaflets were too tall for effective herbicide spraying. Trunks of Japanese angelica trees were cut to permit herbicide treatment of the leaves in 2011.

The record of invasive species adaptive management from the Saddlers's Woods partnership demonstrates the value of assessing strategies in order to effectively achieve the action objectives. Typically, a grid of plots designated as "control" and "treatment" is laid out before treatments occur. Sampling the plots prior to application of the treatment protocol to the "treatment" plots and resampling periodically for at least 2 years enables statistical analysis of treatment effectiveness (Stalter et al. 2009). Regardless of which strategy is being scientifically examined, performance indicators and thresholds for response are needed. Completely effective treatments of invasive species are an unrealistic objective to be avoided. Further, initial successes are not forever because invasive species return and native species plantings die. Therefore, realistic expectations must be set as a level of invasive species decrease and native species increase. Quantifying expectations for results is a partnership

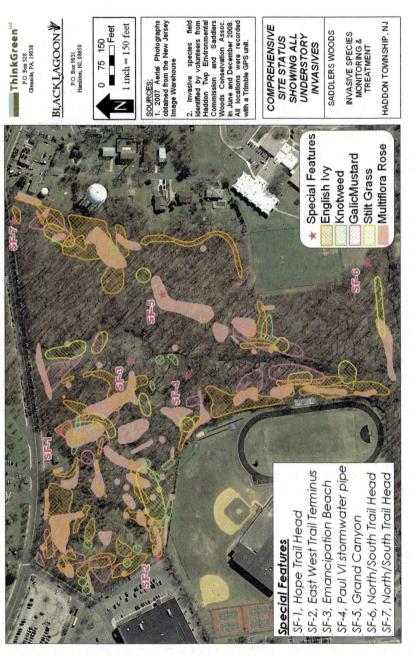

Fig. 3.4 Map of invasive species in Saddler's Woods, Haddon Township, Camden County, New Jersey, in 2008 (Courtesy of the Saddler's Woods Conservation Association, ThinkGreen LLC and Black Lagoon)

decision that should not be driven by public perceptions but rather scientific information and professional experience. Communication about objectives and strategies and their evaluations is essential to build confidence in the partnership.

Every management plan for an old growth urban forest must have an action objective of creating the baseline for assessment, which is a periodic forest canopy, subcanopy, and sapling inventory. When available, identifying tree and sapling locations with GPS and recording locations in GIS is preferable. Annual measurements of seedling populations, abiotic site factors (such as soil, light penetration, and pathways), and threats (including invasive species, herbivores, and off-path trampling) to the forest are needed to monitor forest dynamics (Lindenmayer and Likens 2010). The Green Seattle Partnership has a monitoring methods guide in draft form (Green Seattle Partnership 2010) which is applicable to any forest. Educational institution partners can perform inventories and experimental procedures to assess achievement of management and restoration objectives. However, partnership needs for research and the ability of researchers and students to respond are often quite different. Professors working with graduate students must have projects that can be completed in the course of a year or two at most. Research involving undergraduate students spans a semester unless extended through a summer internship. High-school community service projects provide workers to remove invasive species and plant native species for a few days. Even with well-organized projects, the scope of work can be limited because faculty may not have the equipment needed to conduct some research, but involving interested faculty in writing grants to acquire equipment and supplies from local foundations may help to increase the chance for proposal funding. Although high-school and community college faculty are generally not encouraged to publish in scientific journals, the demands on 4-year college and university professors for scholarly achievements is increasing. Planning the sequence of research, assessment, and restoration projects to fit the available mix of scientists, educators, students, and knowledgeable community neighbors from the partnership requires open discussion so that all of the collaborators are satisfied with the expected research outcomes. A staging approach is often called for to resolve the mismatch between available short-term research capacity and long-term research need. Each project is a semi-discrete entity that is part of a progressive series forming the research program. A project must have sufficiently rigorous objectives and measurable outcomes to satisfy the needs of the participants as well as funding agencies when support is provided for the research. Reevaluating how the available research resources can meet the needs derived from the research-based action objectives is often required to focus the research efforts on new factors affecting the old growth urban forest.

Restoration

Restoration of an old growth urban forest may appear at first glance to be a simple arboricultural process of removing a dead tree and replacing the tree with another one of the same species. The far greater complexity of forest dynamics restoration is made clear by the established guidelines for professional ecological restoration projects which focus on restoring historical continuity at the place of historical occurrence (Clewell et al. 2005). To initiate the planning process, the partnership must first examine what is known about the historical ecology of the old growth urban forest. If proper for the particular old growth urban forest, the next step is to consider questions related to the historical ecology of old growth urban forests in cultures linked to Europe. For example, forest composition transformations occurred in the past as planting preferences changed, which raises the question of whether replacement of the current species in the forest is appropriate. Additional questions arise from the 1818–1819 historical floras information indicating the vast majority of native species were cultivated before the American Revolution: Are the native tree species found in a remnant old growth urban forest actually arboricultural specimens? Should long cultivated alien species be replaced with native species in a street old growth urban forest? What is the value of recreating the historically authentic natural landscape design during the restoration of a landscaped forest if major species such as American elm and chestnut are not available? Do the old growth forests in rural areas serve as a valid guide for species selection in plantings for a remnant old growth urban forest?

The questions derived so far from the historical ecology research are only focused on species composition but historical continuity restoration also considers forest dynamics. Plantings to restore forest dynamics are based on what is needed to rebuild species diversity in the sapling, subcanopy, and canopy layers. The historical continuity restoration plan must assure that the plantings will survive in the street or landscaped forests and engender spontaneous forest regeneration in a remnant forest. Some factors affecting the forest, such as poor soils, can be addressed at the time of plantings. Other problems require change in the relationship of the forest to other creatures. For example, what will protect the plantings from deer browsing and human trampling? If the old growth urban forest restoration objective is spontaneous forest regeneration, then human trampling and deer browsing are the critical problems to address. Urban foresters have repeatedly observed that white-tailed deer are changing urban forest dynamics of the eastern United States (Asnani et al. 2006; Rossell et al. 2007). However, human trampling also changes forest dynamics (Malmivaara-Lämsä et al. 2008) and co-occurs with white-tailed deer browsing. Research in urban forests has not identified the changes in forest dynamics attributed to deer browsing without separating the confounding factor of human trampling. To guide restoration planning, the following research shows the effects of deer browsing with and without human trampling.

Original Research: Walbridge Forest, National Zoological Park, Washington, DC

Walbridge Forest, National Zoological Park, Washington, DC (38° 55′ 57″ N and 77° 2′ 48″ W) was acquired with the original land purchase for the Park in 1889 (Edwin A. Greenough & Co 1889). The tract has a shape similar to a narrow obtuse triangle with Beach Road forming the hypotenuse to the south and the intersection of the remaining two sides of the triangle being the intersection of Klingle Road and Adams Mill Road. The forest is on a steep slope (>25°). Although the forest is located just south of the Piedmont section of the oak-chestnut forest region (Braun 1950), chestnut tree stems per hectare was 25.6 in 1921 as indicated by the number of trees lost to the chestnut blight (Hamlet 1985). Some of the forest and the fence along Beach Road were removed during the Beach Road expansion project in 1967 (Farrell 2004). Development of sewer lines across the Park in 1970 (Hamlet 1985) caused the loss of the old growth urban forest along Adams Mill Road and Klingle Road beyond 240 m from the intersection of both roads. White-tailed deer were first seen in the area of the Park in the 1960s (National Park Service 2009).

In 1955, Dix (1957) sampled the forest using a modification of the random pairs method and collected the dbh for 80 trees and 160 saplings; however no information on transect position within the forest was given. Also, Dix reported the stand was fenced since 1913 and there was no record of fire, grazing, or cutting since 1888. During the reconnaissance of Walbridge Forest in May 2010, white-tailed deer were present. The section of the forest along Adams Mill Road was used as a homeless person encampment with clothes found throughout the area (a hole was cut in the fence about 100 m from the intersection of Kingle and Adams Mill Roads). In sharp contrast, the Kingle Road section had no evidence of use by people. The field work in May 2010 was done in the Kingle Road and Adams Mill Road sections and the method used by Dix (1957) was duplicated except two parallel lines of 20 points each were measured in both sections. The lines and points on the lines were separated from each other by 10 m and were at least 10 m from the border of the forest. Only major species (>10% of saplings, subcanopy trees, or canopy trees in 1955 or 2010) are compared because sample line placements within the forest were different for 1955 and 2010. Seedlings were counted in 1 m by 1 m plots at each point. White-tailed deer were observed each day during the sampling but no people were in the forest.

Based on a comparison of the sapling and tree data, Dix (1957) predicted that American beech and sugar maple would become the dominant species. The 2010 data certainly indicates American beech has increased in the saplings, subcanopy, and canopy layers (Table 3.1). Sugar maple had a larger percentage of the sapling and subcanopy layers in 2010 than 1955 but was not present in the canopy. Beaudet and colleagues (2007) related sugar maple not growing and regenerating as quickly as American beech to the presence of American beech root sprouts, which are common in Walbridge Forest. Although the Klingle Road section had almost twice as many sugar maple saplings as the Adams Mill Road section, American beech showed little difference between the sections for saplings. However, American beech had more than three times as many saplings as sugar maple in both sections. Sugar maples had no leaves below 1.5 m above the ground but leaves and branches were present along the full length of the American beech trunks. The presence of a browse line for sugar maple and not American beech as well as the higher number of American beech saplings than sugar maple saplings could be explained by white-tailed deer preferring to browse sugar maple over American beech (Krueger et al. 2009).

Table 3.1 Percentage of stems for saplings (2.5<dbh<10.0 cm), subcanopy trees (10.0 ≤ dbh ≤ 40.0 cm), and canopy trees (dbh>40.0 cm) for major species (>10% of saplings, subcanopy trees, or canopy trees) and total stems in Walbridge Forest, National Zoological Park, Washington, DC, in 1955 and Adams Mill Road and Klingle Roads section of Walbridge Forest in 2010

	Saplings			Subcanopy			Canopy		
		Adams	Klingle		Adams	Klingle		Adams	Klingle
Species	1955	2010	2010	1955	2010	2010	1955	2010	2010
Acer rubrum	10	6.8	1.25	2.56	3.7	0	0	0	0
Acer saccharum	12.5	10.6	20	2.6	14.8	21.6	0	0	0
Fagus grandifolia	40	66.3	64.4	35.9	50	56.9	12.2	19.2	27.6
Liriodendron tulipifera	1.3	0	0	2.6	1.9	0	12.2	7.7	6.9
Quercus alba	0	0	0	38.5	11.1	0	43.9	42.3	55.7
Quercus rubra	2.5	1.25	1.25	0	3.7	5.88	14.6	15.4	6.9
Quercus velutina	0	0	0	7.69	0	0	14.63	7.69	3.45
Total stems	160	160	160	39	54	51	41	26	29

Table 3.2 T-test (two tailed) of differences between means for maple (*Acer* spp.) and oak (*Quercus* spp.) seedlings present in 40 plots located in the Adams Mill Road (deer browsing and human trampling) and 40 plots in the Klingle Road (only deer browsing) sections of Walbridge Forest, National Zoological Park, Washington, DC, in May 2010

Seedling genus	Forest section	Mean	Standard deviation	T-test significance
Acer	Adams Mill Road	1.45	1.57	>0.05
	Klingle Road	2.35	2.08	
Quercus	Adams Mill Road	0.2	0.52	>0.05
	Klingle Road	4.35	6.55	

In all 80 seedling plots, every maple and oak seedling had cotyledons and the four American beech seedlings were over 50 cm tall. The presence of cotyledons on all of the oaks and maple seedlings indicates seed germination occurred in the prior few weeks and browsing by white-tailed deer is killing all of the oak and maple seedlings. To ascertain the effect of human trampling, the means for the maple and oak seedling counts in the Adams Mill Road plots and Klingle Road plots were compared with T-tests (two-tailed) to determine if there were differences between the two sections of Walbridge Forest (Table 3.2). The T-test calculations were performed using PASW Statistics (formerly SPSS Statistics) version 17 and the significance level selected was 0.05. The means in the Adams Mill Road plots were significantly less than the Klingle Road plots for both oak and maple seedlings, which indicates human trampling reduces seed germination independently of the effects of deer browsing on seedling survival. In conclusion, human trampling reduces seed germination and seedling population size while white-tailed deer are responsible for the loss of oak and maple reproduction but permit American beech seedling and sprouts to survive.

The research in Walbridge Forest revealed the statistically significant differences in the effects of human trampling and deer browsing versus deer browsing on arboreal seedlings and the long-term effects of deer browsing on saplings populations in an old growth urban forest. With the Walbridge Forest results in mind, plans to restore the historical continuity of open public access remnant old growth urban forests in the eastern United States appear to be doomed to fail over the long-term. The restoration of historical continuity in remnant old growth urban forests requires spontaneous regeneration to provide seedlings and saplings of the subcanopy and canopy trees. However, plantings and soil modifications to enhance seedling regeneration (Sullivan et al. 2009) are rendered ineffective by deer browsing with the exception of tree species avoided by deer. In addition, human trampling affects seedlings of all species but to a lesser extent than deer browsing. How does a partnership dodge the 1–2 knockout punch of deer browsing and human trampling when planning for the restoration of a remnant old growth urban forest? The hard, cold answer is: (1) fencing in the remnant old growth forest to exclude deer, (2) building raised pathways with small fences to invite viewing of the forest but not enable trampling by people (Fig. 3.5), and (3) placing ground-level cattle guards at the entrance to the raised pathways in order to block deer access to the pathways.

Going half way, reducing the deer population and building ground-level pathways is at best a way to reduce the rate of tree and sapling losses but will do nothing to reinvigorate spontaneous regeneration. Removing deer raises the American consciousness against hunting deer created by the movie *Bambi* (Lutts 1992), which is an especially strong sentiment for urban people (National Park Service 2009). Even the best methods to create and maintain ground-level trails (Hesselbarth et al. 2007) do

Fig. 3.5 Photograph of raised wooden bridge and raised wood chip trail with low fence built after loss of former South Cove trail to landslides in Radnor Lake State Natural Area, Nashville, Tennessee (Permission of photographer Joshua Walsh)

not inhibit people from walking off the pathway and trampling the forest. Laws with fines for off-pathway transit are only as effective as the law enforcement officers.

Start small with plans for fencing a remnant old growth urban forest. A project to fence a large tract is likely to be resisted because some community members will consider the idea elitist. The first step is raising awareness of how an old growth urban forest is a unique, special, and valuable asset for the community. Next, translate public awareness into the public support essential to implement a small demonstration project through open meetings. Guidelines and action steps for a limited access historical continuity restoration demonstration project including fencing, raised pathways, cattle guards, and plantings are provided in Fig. 3.6.

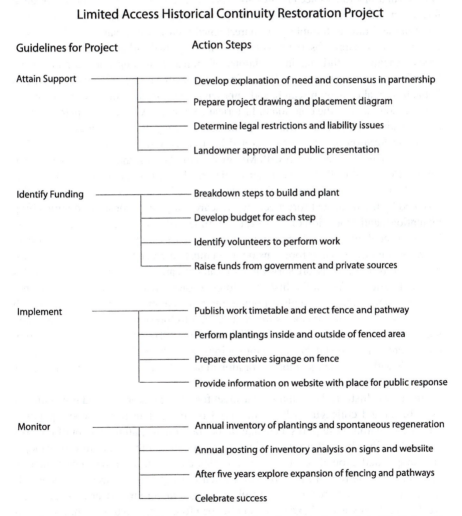

Limited Access Historical Continuity Restoration Project

Guidelines for Project   Action Steps

Attain Support
- Develop explanation of need and consensus in partnership
- Prepare project drawing and placement diagram
- Determine legal restrictions and liability issues
- Landowner approval and public presentation

Identify Funding
- Breakdown steps to build and plant
- Develop budget for each step
- Identify volunteers to perform work
- Raise funds from government and private sources

Implement
- Publish work timetable and erect fence and pathway
- Perform plantings inside and outside of fenced area
- Prepare extensive signage on fence
- Provide information on website with place for public response

Monitor
- Annual inventory of plantings and spontaneous regeneration
- Annual posting of inventory analysis on signs and websiite
- After five years explore expansion of fencing and pathways
- Celebrate success

Fig. 3.6 Implementation guidelines and action steps for a demonstration project of limited-access historical continuity restoration

In no more than 5 years after implementing the project, the differences in plantings survival and spontaneous regeneration between the fenced forest and the surrounding unfenced forests will motivate public interest in expansion of the limited access historical continuity restoration initiative in order to rejuvenate all of the remnant old growth urban forest. Be certain to point out the plantings survival and spontaneous regeneration successes in signs on the fence and in public announcements including the partnership Web site.

The following question may be raised: what is the alternative method to preserve the old growth urban forest? Finding the answer to that question requires an understanding of the concept of shifting baseline syndrome (Pauly 1995). What modern observers know as old growth urban forest is not what late-eighteenth-century observers knew as old growth urban forests because tree species have been lost to disease and the original forest dynamics were destroyed with forest clearance. If the changes caused by deer browsing and human trampling are unrestrained then forest dynamics will be transformed so that future observers will know an old growth urban forest composed of species capable of thriving in conditions of human trampling and deer browsing. Therefore, any alternative method to fencing, raised pathways, planting species revealed in the historical ecology research, and improving growing conditions for spontaneous regeneration, fails on the criterion of historical continuity. A restoration plan based on an alternative method is a schema for planting a currently unknown future forest.

The restoration process for street and landscaped old growth urban forests has many components in common with Miyawaki's method for the restoration of urban forests (Miyawaki 2004). Specifically, historical forest ecology research is performed to define forest composition, dense mixed plantings of the tree species revealed by the historical forest ecology research, topsoil reconstruction, mulching plantations, and protection of the forest over time. All of the species revealed by the historical ecology research may not survive diseases and insect borers or urban growing conditions. Therefore consideration must be given to whether maintenance efforts are available to assure survival or to wait until arboricultural research provides resistant varieties. When the historical forest ecology research is not fruitful, species represented among the older population in the street or landscaped old growth urban forest are better prospects for contributing to forest stability than little-tested species added to increase diversity (Richards 1983). Tree removals for restoration efforts are controversial but communications with the public must associate the removals with the benefits of the restoration plan goals in order to resolve objections (Weinstein 1984).

To achieve historical continuity in a street forest, the restoration planting concept as Sullivan and colleagues (2009) observed is "more than filling space quickly." Historical continuity requires a change in the traditional spatial concept of one tree in one location. Instead a series of microforests containing a mixture of canopy, subcanopy, and ground layer species are planted. All three forest layers can be placed in a single row or in double rows when spacing between trees provides sufficient room for each tree to grow a full crown as a mature tree (Carmichael 1995). Replacing street soils and exhausted landscaped forest soils to the extent possible is essential. Maintenance efforts, particularly providing water to the trees must be continued for years to assure that the plantings will become established. The planting of microforests contributes to reducing maintenance and replacement costs because trees in groups serve as barriers to trampling (Lehvävirta 1999; Lehvävirta et al. 2004). The research methods described above for creating a forest inventory to serve as a baseline for monitoring the forest are also essential for street and landscaped forests in order to enable adaptive management responses to failures and successes in tree plantings and maintenance. Finally, the historical continuity restoration plan for a street or landscaped old growth urban forest must set a goal and formulate objectives for ongoing plantings to create the substitute for spontaneous regeneration in a remnant old growth urban forest.

In summary, restoration projects for old growth urban forests face many challenges to success including tree losses related to abiotic and biotic factors such as the people interacting with the forest. Directly addressing the threats to the canopy, subcanopy, and sapling layers of the old growth urban forest through activities such as invasive species removal and native species plantings is only a part of what the partnership needs to do to restore the historical continuity of the forest. Communications with the public and education of the young are required to change perceptions of the importance of an old growth urban forest restoration project. For remnant old growth urban forests with little or no spontaneous regeneration because of human trampling and deer browsing, a model process for implementing a raised pathway and fenced forest demonstration project is given to enable the survival of the seedling and sapling populations that are essential for restoration of historical continuity. A second model for historical continuity of street and landscaped old growth urban forests is planting microforests in order to provide the canopy, subcanopy, and ground layers of a forest and planning for replacement of trees over the long-term. The final step is yours, identify old growth urban forests in your metropolis, perform the historical ecology research, form partnerships, and undertake the processes to restore the historical continuity of the old growth urban forest.

References

Asnani KM, Klips RA, Curtis PS (2006) Regeneration of woodland vegetation after deer browsing in Sharon Woods Metro Park, Franklin County, Ohio. Ohio J Sci 3:86–92
Beaudet M, Brisson J, Gravel D, Messier C (2007) Effect of a major canopy disturbance on the coexistence of *Acer saccharum* and *Fagus grandifolia* in the understorey of an old-growth forest. J Ecol doi:10.1111/j.1365-2745.2007.01219.x
Braun EL (1950) Deciduous forests of eastern North America. Hafner, New York, NY
Carmichael R (1995) Avenues from the past to the future. Arboric J 19:111–120
Clewell A, Rieger JP, Munro J (2005) Guidelines for developing and managing ecological restoration projects. Soc Ecol Restor Int, Tucson, Ar
Dix RL (1957) Sugar maple in forest succession at Washington, DC. Ecol 38:663–665
Edwin A. Greenough & Co (1889) Map of Zoological Park showing boundary lines and property lines in the park. United States Geological Survey, Washington, DC
Elmendorf WF, Willits FK, Sasidharan V, Godbey G (2005) Urban park and forest participation and landscape preference: a comparison between blacks and whites in Philadelphia and Atlanta, U.S. Arboric Urb For 31:318–326

Farrell G (2004) Smithsonian Institution National Zoological Park: a historic resource analysis. Unpubl, Smithsonian Institution, Washington, DC

Gobster PH (1997) The Chicago wilderness and its critics III. The other side: a survey of the arguments. Restor Manag Notes 15:32–37

Green Seattle Partnership (2010) Monitoring data collection methods. http://greenseattle.org/forest-steward-resources-1/monitoring/draft-monitoring-protocols. Accessed 6 April 2011

Green Seattle Partnership (2011a) Tree ambassador program. http://seattle.gov/trees/treeambassador.htm. Accessed 6 April 2011

Green Seattle Partnership (2011b) Forest steward field guide. http://greenseattle.org/forest-steward-resources-1/new-forest-steward-field-guide-2011-1/at_download/file. Accessed 6 April 2011

Hamlet SE (1985) The National Zoological Park from its beginnings to 1973. Unpubl, National Zoological Park, Washington, DC

Hesselbarth W, Vachowski B, Davies MA (2007) Trail construction and maintenance notebook: 2007 edition. Tech. Rep. 0723-2806-MTDC. US Dep Agric, For Serv, Missoula Tech Dev Cent, Missoula, Mt

Jones N, Collins K, Vaughn J, Benedikz T, Brosnan J (2005) The role of partnerships in urban forestry. In: Konijnendijk CC, Nilsson K, Randrup TB, Schipperijn J (eds) Urban forests and trees: a reference book. Springer, Berlin, Ger

Krueger LM, Peterson CJ, Royo A, Carson WP (2009) Evaluating relationships among tree growth rate, shade tolerance, and browse tolerance following disturbance in an eastern deciduous forest. Can J For Res doi:10.1139/X09-155

Lehvävirta S (1999) Structural elements as barriers against wear in urban woodlands. Urb Ecosys doi:10.1023/A:1009513603306

Lehvävirta S, Rita H (2002) Natural regeneration of trees in urban woodlands. J Veg Sci doi:10.1111/j.1654-1103.2002.tb02023.x

Lehvävirta S, Rita H, Koivula M (2004) Barriers against wear affect the spatial distribution of tree saplings in urban woodlands. Urb For Urb Green doi:10.1016/j.ufug.2003.10.001

Lindenmayer DB, Likens GE (2010) Effective ecological monitoring. Earthscan, Washington, DC

Linnea A (2010) Keepers of the trees: a guide to the re-greening of North America. Skyhorse Publishing, New York, NY

Loeb RE (1987) The tragedy of the commons: an update – can urban foresters save city parks? J For 84:28–33

Loeb RE (2009) Biogeography of invasive plant species in urban park forests. In: Kohli R, Jose S, Batish D, Singh H (eds) Invasive plants and forest ecosystems, CRC/Taylor and Francis, London, UK

Loeb RE, Germeraad J, Treece T, Wakefield D, Ward S (2010) Effects of one-year versus annual treatment of Amur honeysuckle in forests. Invasive Plant Sci Manage 3:334–339

Lutts RH (1992) The trouble with Bambi: Walt Disney's Bambi and the American vision of nature. For Conserv His 36:160–171

Maitland MC (1995) Squirrel strategy. Arboric J 19:349–356

Malmivaara-Lämsä M, Hamberg L, Löfström I, Vanha-Majamaa I, Niemelä J (2008) Trampling tolerance of understorey vegetation in different hemiboreal urban forest site types in Finland. Urb Ecosys doi: 10.1007/s11252-007-0046-3

McFadden JE, Hiller TL, Tyre AJ (2011) Evaluating the efficacy of adaptive management approaches: is there a formula for success? J Environ Manage doi:10.1016/j.jenvman.2010.10.038

Millward AA, Paudel K, Briggs SE (2011) Naturalization as a strategy for improving soil physical characteristics in a forested urban park. Urb Ecosyst doi:10.1007/s11252-010-0153-4

Miyawaki A (2004) Restoration of living environment based on vegetation ecology: theory and practice. Ecol Res doi:10.1111/j.1440-1703.2003.00606.x

Müller N, Werner P (2010) Urban biodiversity and the case for implementing the convention on biological diversity in towns and cities. In: Müller N, Werner P, Kelcey JG (eds) Urban biodiversity and design. Blackwell Publishing Ltd, Oxford, UK

National Park Service (2009) Draft white-tailed deer management plan environmental impact statement Rock Creek Park, Washington, DC. Unpubl, National Park Service, Washington, DC

Pauly D (1995) Anecdotes and the shifting baseline syndrome of fisheries. Trends Ecol Environ 10:430

Profous G, Loeb RE (1990) The legal protection of urban trees: a comparative world survey. J Environ Law 2:179–193

Raffetto J (1993) Perceptions of ecological restorations in urban parks. Policy recommendations and directions: a Lincoln Park case study. In: Gobster PH (ed) Managing urban and high-use recreation settings: selected papers from the urban forestry and ethnic minorities and the environment paper sessions at the 4th North American symposium on society and resource management; May 17–20, 1992 University of Wisconsin Madison, Wisconsin. Gen Tech Rep NC-163, US Dep Agric, For Serv, North Cent For Exp Stn, Chicago Il

Richards NA (1983) Diversity and stability in a street tree population. Urb Ecol 7:159–171

Rossell CR, Patch S, Salmons S (2007) Effects of deer browsing on native and non-native vegetation in a mixed oak-beech forest on the Atlantic coastal plain. Northeast Nat doi:10.1656/1092-6194(2007)14[61:EODBON]2.0.CO;2

Stalter R, Kincaid D, Byer M (2009) Control of nonnative invasive woody plant species at Jamaica Bay Wildlife Refuge, New York City. Arboric Urb For 35:152–156

Stein AB, Moxley JC (1992) In defense of the nonnative: the case of the eucalyptus. Landsc J 11:35–50

Sullivan JJ, Meuk C, Whaley KJ, Simcock R (2009) Restoring native ecosystems in urban Auckland: urban soils, isolation and weeds as impediments to forest establishment. New Zealand J Ecol 33:60–71

Tu M, Meyers-Rice B (2001) Site weed management plan template. Nature Conservancy. http://www.invasive.org/gist/products.html. Accessed 18 February 2011

Van Hertzele A, Collins K, Tyrväinen L (2005) Involving people in urban forestry – a discussion of participatory practices throughout Europe. In: Konijnendijk CC, Nilsson K, Randrup TB, Schipperijn J (eds) Urban forests and trees: a reference book. Springer, Berlin, Ger

Vidra RL, Shear TH, Stucky JM (2007) Effects of vegetation removal on native understory recovery in an exotic-rich urban forest. J Torrey Bot Soc doi:10.3159/1095-5674(2007)134[410:EOVRON]2.0.CO;2

Weinstein G (1984) Central Park, New York-a management survey of an urban forest. Arboric J 8:321–330

Westphal LM, Childs GM (1994) Overcoming obstacles: creating volunteer partnerships. J For 92(10):28–32

Zipperer WC (2010) Factors influencing non-native trees species distribution in urban landscapes. In: Müller N, Werner P, Kelcey JG (eds) Urban biodiversity and design. Blackwell Publishing Ltd, Oxford, UK

Index